GLOBAL GEOPHYSICS

General Editor

SIR GRAHAM SUTTON, C.B.E., D.Sc., LL.D., F.R.S.

Physics of Lightning
D. J. MALAN, D.Sc., F.R.S.S.Af.

The English Climate
H. H. LAMBE, M.A.

Scientific Method
M. WEATHERALL, M.A., D.M., D.Sc.

Observation in Modern Astronomy
D. S. EVANS

ALSO OF INTEREST

Magnetic Domains
and Techniques for their Observation
R. CAREY, B.Sc., Ph.D.
E. D. ISAAC, B.Sc., Ph.D.

Principles of Physical Geography
F. J. MONKHOUSE, M.A., D.Sc.

GLOBAL GEOPHYSICS

R. H. TUCKER, M.Sc., F.R.A.S.
A. H. COOK, Sc.D., F.R.S., F.R.S.E.
H. M. IYER, Ph.D.
F. D. STACEY, Ph.D., D.Sc., F.A.I.P.

AMERICAN ELSEVIER PUBLISHING COMPANY, INC.
New York 1970

Published in the United States by
AMERICAN ELSEVIER PUBLISHING COMPANY, INC.
52 Vanderbilt Avenue
New York, N.Y. 10017

First printed 1970

Library of Congress Catalog Card Number: 78-122635
Standard Book Number: 444-19648-X

Printed and bound in Great Britain

Editor's Introduction

Although the study of the Earth as a physical system goes back to the beginnings of western science, the greatest rate of advance in our knowledge has occurred in the last few decades. Electronics has made it possible to observe and measure features of the solid Earth, its atmosphere and oceans, on a scale and with an accuracy and precision never before attained. Equally important, the invention of the high-speed computer has enabled the vast amounts of data so obtained to be processed and analysed expeditiously, so that, all told, the prospect for the earth sciences is now bright. To-day the younger generation of mathematicians and physicists is showing a lively curiosity about subjects that in the last generation were regarded as somewhat esoteric, the chosen field of a few devotees.

In its strict sense, the word 'geophysics' covers the study of the whole Earth, from its inaccessible fiery interior to the remote fringe of its atmosphere. In current practice, it is used in two rather different specialized senses. For professional geologists, especially those concerned with the tasks of survey or prospecting, it usually denotes a group of techniques developed to help in elucidating the nature and structure of limited regions of the Earth's crust. For others it implies the study of the solid Earth as a whole, and it is in this sense that the word is used here. There is no dearth of accounts of global geophysics but they are, for the most part, highly mathematical treatises, and it is often a daunting task for the newcomer to obtain from them a broad understanding of the subject. There are also those, especially students of geography, who find the standard treatises beyond their scope because they lack the necessary long preliminary training in mathematics and physics.

This book has been planned as an exercise in the communication of ideas, a bridge between the condensed articles of the scientific encyclopedias and the detailed expositions of the professional texts. The Prologue, which sets the scene against the astronomical background, is part of a much longer essay by Mr. Tucker which, unfortunately, has had to be curtailed here for reasons of space. The remainder of the book deals at greater length with three major disciplines: geodesy, seismology and geomagnetism. Professor Cook describes investigations on the Earth as a solid rotating body with a powerful gravitational field. For Dr. Iyer, the Earth is an elastic body, subject to tremors ranging from microseisms to devastating earthquakes. Finally, Dr. Stacey considers the Earth as a magnetic body with surrounding radiation belts.

In the planning of this book the authors, all of whom are active workers in their fields, were given complete freedom in the choice of material and the method of presentation, with the suggestion that they should try to keep the mathematics to a minimum. In some subjects, notably geodesy, this is extremely difficult; others, such as seismology, allow a more descriptive approach. There has been no attempt to evade difficult ideas, but from time to time the reader is asked to accept some statement on trust. This is unlikely to impair the value of the book for the serious reader, for the details can always be found in the literature. The underlying philosophy has been that, in surveys of this kind, ideas matter more than proofs.

Observations made during the International Geophysical Year and those now being received in enormous quantity from artificial satellites have given a great impetus to the study of global geophysics. It is my hope that this book will be welcomed in schools and universities as a not too difficult introduction to what is known and being discovered about certain aspects of the physical nature of our planet. In particular, I hope it will excite enough interest in the minds of young people to persuade some of them that global geophysics is sufficiently attractive and rewarding to be made a lifetime study.

O.G.S.

Contents

Prologue: The Earth in Space

R. H. Tucker, Royal Greenwich Observatory, Herstmonceux

1.1 The Solar System

'The Earth is a planet, and belongs to the Solar System', so we are told, but the statement does not mean very much until we know what a planet is, and can say what we mean by the Solar System. The modern meaning of the word 'planet' is: 'a relatively small, cool body moving around a large, hot body (a star) under its gravitational influence'. Strictly speaking, this is the definition of a *primary planet*; it is possible for a primary planet to have even smaller bodies, known as *secondary planets*, or *satellites*, moving around it. For example, the Sun is the large hot body, or star, around which the small cool planet Earth is moving, while the Moon is a smaller body moving around the Earth, and is therefore the Earth's satellite.

The Earth is not the only planet moving around the Sun. There are at least eight other major planets, some of them accompanied by satellites, and several thousand *minor planets* or *asteroids*, ranging in size up to a few hundred miles in diameter, each moving around the Sun in its own path or orbit. In addition, there are millions of even smaller objects in orbit around the Sun, the *meteoroids*, which are not visible until they enter the Earth's atmosphere, where they become heated by air friction and appear as meteors or shooting stars. Finally, there are the spectacular bodies called *comets*, which apparently consist of a group or cluster of meteoroids surrounded by a cloud of glowing gas, journeying in company around the Sun, and often possessing a magnificent tail stretching out through space on the side away from the Sun.

All these various classes of objects have one thing in common, their membership of the Solar System. They all remain in the vicinity of the Sun, under its dominating gravitational influence, and receive their light and heat from it. They form a relatively compact gathering around the Sun, surrounded on all sides by large expanses of nearly empty space. If we could travel out through this space with the speed of light, it would be several years before we reached the next star, which gives an idea of the emptiness of interstellar space. If we continued travelling through the stellar system for thousands of years, we should eventually find that the stars were thinning out, and it would become clear that they were grouped into a huge disk-like shape. This disk, known as the *galaxy*, contains about a hundred thousand million stars. Outside the galaxy we find more expanses of empty space, separating our galaxy from other galaxies, and there are millions of galaxies stretching away into the furthest depths of space.

The Solar System consists of the Sun, the nine major planets and their satellites, the minor planets, the meteoroids and the comets. Outside the Solar System is the rest of the universe, comprising the stars of our galaxy, and all the other galaxies.

The Sun, the central dominating body of the Solar System, the source of its light and heat, is in fact a very ordinary star, one of millions, rather below the average in size and brightness. Its supremacy in the sky is due to its nearness to us, and to the remoteness of its nearest rivals. From the cosmic viewpoint there is nothing very special about the Sun, and for all we know there is nothing very special about the Solar System. Even if only one star in a million possesses a family of attendant planets, there will be something like a hundred thousand such families in our own galaxy alone, to say nothing of what may be in the millions of other galaxies. The only way in which our Solar System is unique is that we are in it and close enough to observe many of its members. We see the Sun as a glowing ball, occasionally marked by spots, while we see the stars as sharp points of light in which no details can be discerned. Yet the Sun and the stars are very much alike. We see the major planets as other points of light, resembling the stars, but in fact the planets are as different from the stars as cricket balls are from blast-furnaces. We see the Moon as a crescent or a disk, sometimes bright enough to rival the Sun, but in fact the Moon and Sun are as different as planets and stars. It is one of the fascinations of astronomy that things are not what they seem, and the history of that science is largely the history of the liberation of the imagination. We have come to realize that the apparently complex happenings in the sky are really fairly straightforward happenings in space as seen from our viewpoint.

Once the Sun is accepted as an ordinary star, we can apply to it the theories of evolution of stars that have been built up on the study of the stellar universe. We can trace the possible history of the Sun back to a large body of cool gas which somehow managed to separate itself from a larger cloud of gas. The body of gas then began to shrink under its own gravitation, causing its temperature to rise until the centre was hot enough for the hydrogen atoms to start a thermo-nuclear reaction. There was now energy to spare, and the gas would have to get rid of it by radiating it away into space. So the Sun began to shine, as it has been shining for millions of years, apparently with surprisingly little variation. We cannot be so confident about the early history of the planets, mainly because we have not been able to see the planets that other stars may possess, and so we cannot construct an evolutionary sequence of planetary systems in the way that we can for stars themselves. It could be that the planets were formed during the initial shrinking of the solar gas, in which case the planets are a little older than the Sun. Another possibility is that the planetary material was torn out of the Sun by some catastrophe at a later stage, in which case the planets are younger than the Sun. The only certainty seems to be that the planets and the Sun were made out of the same sort of material.

How, then, are we to account for the vast difference between the Sun and the planets as we see them today? The reason must lie in the great disparity of size and mass. The most massive planet, Jupiter, has less than one thousandth part of the Sun's mass, and if we add all the masses of the planets together we can make up only about one part in 750 of the mass of the Sun. It is to be expected, therefore, that even if the Sun and the planets were of identical constitution when first formed there would be a tendency for them to develop in very different ways. One important difference would be in the rate of cooling; small bodies have relatively greater surface areas than large bodies, and therefore cool more quickly. Again, small bodies cannot develop as high a central pressure as large bodies, and so it seems that there is no possibility of starting up a thermo-nuclear reaction by self-gravitation in masses smaller than a certain critical value. Yet another difference arising from inequality of size and mass is in the gravitational attraction at the surface of the body, which affects the ability of the body to retain the lighter gases, such as hydrogen and helium. Consequently we find that the Sun is composed of hydrogen and helium, with a small proportion of heavier elements, while the Earth has lost most of its light gases, leaving a preponderance of the heavier elements. The Moon, being even smaller and less massive than the Earth, has lost the whole of its atmospheric gases; while Jupiter, with a mass over 300 times that of the Earth has an atmosphere consisting almost entirely of hydrogen in various forms.

One of the important features of the Solar System is that it comprises bodies with very differing surrounding atmospheres. The Sun is entirely gaseous in character, but all the sunlight comes from a fairly shallow layer called the photosphere beyond which we cannot see, so we usually regard this as the surface of the Sun, and refer to the layers of gas above this surface as the Sun's 'atmosphere'. The planets of large mass, Jupiter, Saturn, Uranus and Neptune, have extensive atmospheres of hydrogen and its compounds, ammonia and methane. The planets of moderate mass, Earth and Venus, have moderate atmospheres which lack hydrogen. The planets of small mass, Mars and Mercury, have little or no atmosphere. Some of the larger satellites of the more remote planets may have succeeded in retaining an atmosphere, because retention is easier at the lower temperatures which are prevalent at those greater distances from the Sun, but the general rule is that satellites have no atmospheres. It has not yet been possible to determine the mass of Pluto accurately, but it may well be a planet of moderate mass like the Earth and Venus, and so able to retain an atmosphere of the medium and heavy gases.

Another striking feature of the principal members of the Solar System is the fact that all the planets travel round the Sun in the same direction, and nearly in the same plane. The Solar System is therefore very 'flat'; at least, it was so regarded until 1930, when Pluto was discovered. We now know that Pluto moves in a plane which is rather more slanting than is usual for a planetary orbit. The paths or orbits of the planets are more or less circular, being ellipses of small eccentricity except in the cases of Mercury and Pluto, which have orbits of moderately large eccentricity.

At this point is it customary to illustrate the extent of the Solar System by constructing an imaginary scale model, so let us start with a familiar representation of the Earth, a globe one inch in diameter, such as may be found on a certain type of pencil sharpener. To represent the Sun, which

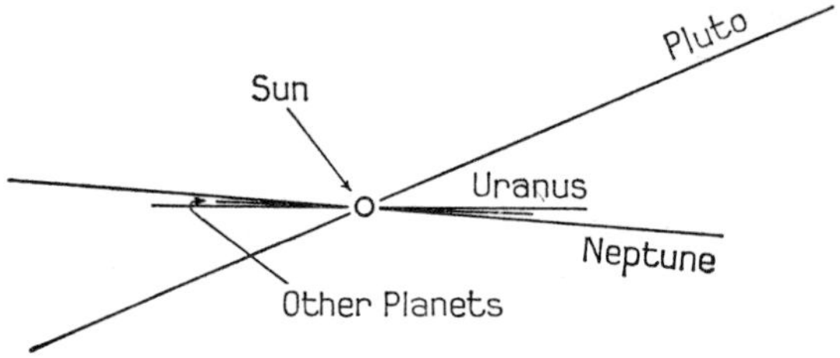

FIG 1 Planetary orbits—apart from Pluto, a fairly 'flat' system

has a diameter 108 times that of the Earth, we must have a globe or ball nine feet in diameter. Having placed our model Sun in a central position, we need to place our model Earth at the correct scale distance from it, which is 328 yards, representing the 93 million miles between the actual Sun and the actual Earth. The next task is to represent and locate the other major planets, so we start with Mercury, the nearest planet to the Sun, represented by a pea placed 127 yards from the model sun. Venus, nearly the same size as the Earth, is a 1 inch ball at 237 yards from the model Sun, and next in order comes the Earth itself, which we have already set in place at 328 yards. Mars, just over half the diameter of the Earth, is a marble at a distance of 500 yards, and the next planet, Jupiter, is an eleven-inch ball at 1,705 yards, which is nearly a mile. Saturn comes next, a small football $9\frac{1}{2}$ inches in diameter at a distance of $1\frac{3}{4}$ miles, followed by Uranus, $3\frac{2}{3}$ inches in diameter at $3\frac{1}{2}$ miles, Neptune, $3\frac{1}{2}$ inches in diameter at $5\frac{1}{2}$ miles, and Pluto, probably about the same size as the one-inch Earth, at an average distance of $7\frac{1}{3}$ miles.

So far the model has given some idea of the 'emptiness' of the Solar System, but it is rather misleading because it gives the impression that the planets have all been set in a straight line. The planets, of course, move around the Sun in their orbits or paths, and so we need even more space for our model if it is to be a working model. Instead of a line seven miles long, we need a circular area over fourteen miles across for the model planets to move in, and they must all be moving around the model Sun with various speeds. The speed of the actual Earth in its orbit is about 19 miles per second, which carries it once around its orbit (radius 93 million miles) in a period of one year. The diameter of the actual Earth is 7,927 miles, and so the Earth requires nearly 7 minutes to traverse its own diameter along its orbit. In our scale model, therefore, we should have to wait 7 minutes for the 1 inch Earth to travel 1 inch along its orbit. The fastest moving planet, Mercury, would take only 4½ minutes to travel 1 inch, and less than 2 minutes to traverse its own diameter of two-fifths of an inch. The slowest planet, Pluto, crawling along on the boundary of our model, would take 45 minutes to travel 1 inch, and would thus take nearly 248 years to complete one circuit around the Sun. In the course of this circuit, Pluto will approach to within 5½ miles of the Sun, and then will swing out

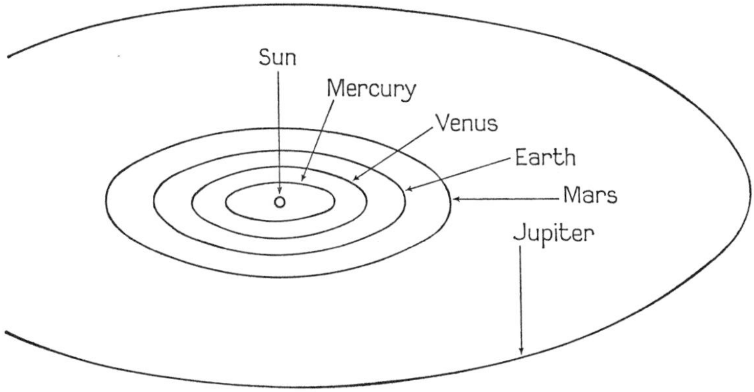

FIG 2 The nearer orbits. Note the 'gap' between Mars and Jupiter

to a greatest distance of over 9 miles, which shows that the orbit is centred nearly 2 miles away from the Sun. At its closest approach to the Sun, Pluto is just able to get closer than Neptune, which then, for a short time, becomes the planet farthest from the Sun.

For simplicity we have assumed that all the planetary orbits are in the same plane, and laid out our model on flat ground. We shall need to modify our model to show the inclination of the various planetary orbits to one another. If we leave the orbit of the Earth on flat ground, to serve as a reference plane, we must then incline the orbit of Mercury by 7 degrees, so that the planet can climb about fifty feet above the level of the Earth's orbit, and then descend 50 feet below it at the other side of the Sun. A similar allowance of height and depth will accommodate the orbits of Venus and Mars. The more extensive orbits of Jupiter, Saturn and Uranus allow them to reach greater heights and depths, amounting to a few hundred feet, and Neptune, at its distance of 5½ miles, ranges nearly a thousand feet above and below the reference plane. But all these are dwarfed by the performance of Pluto, which combines the greatest distance with the greatest inclination, and so can soar over two miles above the plane of the Earth's orbit, and dive to a corresponding depth below it. Thus we see that our model covers not only a large area but also a considerable volume of space. We could regard the Solar System as reasonably flat so long as we consider only the major planets excluding Pluto, but when we add Pluto to the picture, and even more when we include the minor planets and the comets, we are forced to admit that the System has extension in three dimensions and not merely two.

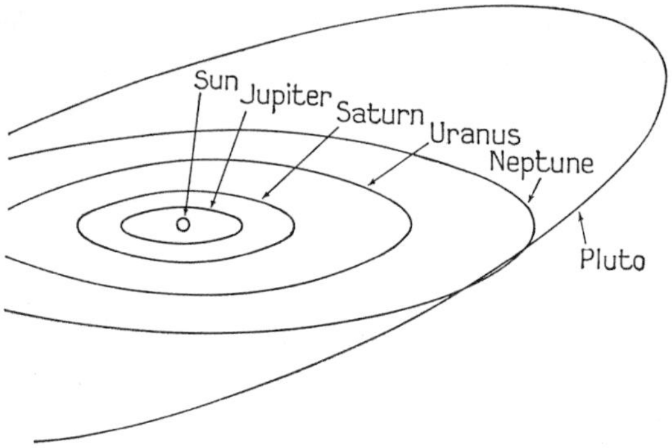

FIG 3 The outer orbits. Pluto can just get 'inside' Neptune's orbit

Before we leave the realm of the major planets, we must mention their subordinate companions or satellites. The best known example is the Moon, which revolves around the Earth in a period of one month. To represent the Moon in our scale model of the Solar System, we take a small pea (a quarter of an inch in diameter) and place it 30 inches from the 1 inch model Earth. Mercury and Venus have no known satellites. Until 1957, the Earth had one, the Moon, but now it has many small artificial satellites revolving close to its surface. Mars has two small, close satellites, while Jupiter has four large satellites and at least eight small ones. Nine satellites of Saturn have been discovered, five of Uranus, and two of Neptune. Most of the satellites revolve about their local planet near the equatorial plane of that planet, and in the same direction as the planet's rotation, but there are a few exceptions, which may arise from a planet capturing a large passing meteoroid and converting it into a satellite. It has been suggested that Pluto was at one time a satellite of Neptune, and that it somehow managed to escape into a nearly separate orbit. The rings of Saturn, which appear to be solid flat structures encircling the planet, are known to consist of millions of small rocks, each constituting a satellite of Saturn.

If we consider the distances of the major planets from the Sun, they seem to increase fairly regularly, each planet being $1\frac{1}{2}$ to 2 times as far from the

Sun as the previous one, but with a noticeable 'gap' between Mars and Jupiter, where the ratio of distances is $3\frac{1}{2}$ to 1. It almost seems as if a planet has been missed out at this point of the sequence, and several astronomers searched the sky in hopes of finding a new planet to fill the gap. At last, on 1st. January, 1801, the astronomer Piazzi discovered a new planet which seemed to fit into place. The new planet was much fainter than the other nearby planets, hardly visible to the unaided eye, and its evident small size and mass were rather puzzling. Within a few years, however, three more planets of the same type were discovered, all having orbits between the orbits of Mars and Jupiter. Later, more and more of these small planets were discovered, until today the number has reached several thousand. They are known as *minor planets*, or *asteroids*, and are usually regarded as the fragments of a major planet which was broken up by some catastrophe. The small mass of a minor planet makes it very vulnerable to the gravitational influence of the other members of the Solar System, and it can easily be swung off its previous path if it happens to get near to another planet. The minor planet orbits are therefore in some disarray, with all sorts of eccentricities and inclinations. There is at least one minor planet which gets nearer to the Sun than Mercury, and another gets further out than Saturn, on an orbit which is very highly inclined (at 42°) to the Earth's orbit.

In contrast with the inconspicuous minor planets, discovered only last century, we have the *comets*, which constitute the most spectacular members of the Solar System, and have been regarded with awe and superstitious fear since the earliest times. A bright comet appears as a rather hazy star, usually provided with a long 'tail' streaming away across the sky, like a luminous cloud of smoke. Conspicuous comets are rather rare, averaging only one in every four or five years, but telescopes reveal that there are many faint comets in the Solar System. The orbits of the comets are in even greater disarray than those of the minor planets, and can have almost any inclination and eccentricity. The comets are the true boundary markers of the Solar System, and form a kind of spherical cloud surrounding the Sun. Any particular individual comet spends most of its time in the outer part of this cloud, and periodically moves inward to complete a circuit of the Sun. When close to the Sun, the comet grows brighter and develops a tail, but it soon moves out again to the more peaceful life of the outskirts, far beyond our sight. In spite of their awe-inspiring appearance, comets are quite harmless objects. The Earth has passed through a comet's tail without any effect being discernible, and it seems that a direct collision with the head of a comet would merely result in an unusually brilliant shower of shooting stars, because the head is in fact only a collection of small rocks and stones journeying in company around the Sun. If a comet came too close to the Earth, the company would be broken up, and the separate stones would pursue individual orbits. The Earth would in fact destroy the comet, and there seems to have been a case of a comet being broken up by planetary action just over 100 years ago. The comet was seen no more, but a shower of shooting stars can be seen when the Earth passes near the orbit of the comet.

This would suggest that shooting stars, or meteors, are individual stones moving in orbits around the Sun, and this qualifies them to be regarded as members of the Solar System. Furthermore, if we admit them as members, they are certainly the most numerous class of members, for it has been estimated that millions of them are swept up by the Earth every day. Most of these are extremely small, and float gently down through the atmosphere, contributing to the surface dust on land, and to the layers of sediment on the ocean beds. It is rather doubtful whether these tiny specks of dust, the micro-meteorites, can be regarded as true members of the Solar System, because their small mass puts them completely at the mercy of any disturbing force from a larger object, and many of them must be swung into the Sun, or into a planetary surface, before they can complete a circuit of the Sun. The continued existence of these small bodies in the region of the Solar System seems to imply that the supply is being replenished from outside in some way. It could well be that the Solar System is passing through a cloud of dust, and capturing some of the particles to serve as temporary members of the System. Indeed, it has been suggested that the whole Solar System, Sun, planets, satellites, asteroids, comets and meteoroids, came into existence by a capturing process of this kind.

1.2 The Spinning Planet

The Earth is not stationary, but executes a rather complicated motion which may be divided into three main components: rotation, revolution and translation. The rotation is the turning of the planet on its axis once per day, the revolution is its orbital motion around the Sun once per year, and the translation is the motion of the whole Solar System through space.

The most obvious effect of the Earth's rotation is the succession of day and night which affords the natural division of time into working and resting periods. Accompanying the alternation of light and dark is an alternation of temperature, with important effects on most forms of life, and with equally important effects on rocks and soil, which are weathered and broken up by the cycle of frost and thaw. These phenomena recur at every rotation of the Earth with respect to the Sun, once in every solar day, being associated with the supply of light and heat from the Sun to the sunlit side of the Earth.

Although the Moon is much smaller and less massive than the Sun, its comparative nearness to the Earth causes it to have a significant gravitational effect on our planet. The Moon attracts the nearer side of the Earth more than the centre, and the centre more than the far side, so that there is a force tending to distort the planet's shape, to make it stretch in the direction toward the Moon, and in the opposite direction, away from the Moon. The waters of the oceans respond to this force and try to accumulate at the point nearest to the Moon, and at the diametrically opposite point of the Earth's surface. The rotation of the Earth drags the water on a little way past the points of accumulation, but there are still two bulges of water, and as a particular point of the Earth is carried past the bulges, the level of the sea rises and falls, and we have the familiar effect of the marine tides. The cycle is governed by the rotation of the Earth with respect to the Moon, and there are two high tides in each lunar day of about 24 hours 50 minutes. The greater length of the lunar day is a consequence of the orbital motion of the Moon around the Earth in the same direction as the Earth's rotation on its axis. While the Earth is completing one rotation with respect to the Sun, the Moon has advanced some way along its orbit, and so the Earth has to turn by an extra amount in order to complete one rotation with respect to the Moon. The influence of the Sun has also to be taken into account in tidal prediction. In addition to the oceanic tides, there are tides in the atmosphere in which a small lunar effect has been found and in the Earth's crust, which moves up and down by a few inches twice in each lunar day (the so-called 'earth tides').

Finally, there are certain effects of the Earth's rotation which do not involve reference to the Sun or Moon, arising from the Earth's rotation in space, with respect to the rest of the universe. These effects include the diminution of the Earth's apparent gravity as we approach the Equator, and the Coriolis force experienced by any object moving across the Earth's surface, which can both be explained as the result of making measurements on a rotating planet while trying to ignore or forget the fact of its rotation. If an observer on the Equator were to ignore the rotation of the Earth and consider himself at rest, on measuring the acceleration due to gravity he could find it smaller than he expected. But he and his laboratory are being accelerated downwards, towards the centre of the Earth, by the rotation of the Earth. This acceleration should be added on to his measured value to obtain the correct total acceleration due to gravity. A similar correction is required anywhere on the Earth's surface, except at the poles, but there is the added complication that for observers not on the Equator the acceleration due to rotation is not vertically downwards, so that the gravity measured is 'wrong' in both magnitude and direction. (Perhaps it is unfair to describe the measured value as 'wrong', since it is perfectly valid for all practical purposes in that locality. Its only imperfection is that it does not truly represent the purely gravitational attraction of the Earth.)

The diminution of the apparent gravity is sometimes explained in terms of a 'centrifugal force' which tends to throw 'stationary' objects on the Earth's surface away from the axis of rotation: the force which appears to affect moving objects, the Coriolis force, is of the same nature, being a fictitious force introduced to explain the result of forgetting or ignoring the Earth's rotation. In the well-known Foucault pendulum experiment, a massive body is freely suspended by a long wire and allowed to swing to and fro with the least possible outside interference. In the northern hemisphere the plane in which the pendulum swings shows a tendency to rotate clockwise (from north to east, south and west), and the rate of rotation increases with north latitude. In the southern hemisphere the rotation is in the other direction, and on the Equator there is no tendency for the plane to rotate at all. The direct explanation is that the Earth is rotating under the pendulum: the rotation can be separated into a rotation about the observer's vertical, which the pendulum can reveal, and a rotation about the observer's north-south horizontal line, which does not affect the pendulum. An alternative explanation, which dispenses with the rotation of the Earth, is that the moving bob of the pendulum experiences a force proportional to its velocity at any time which tends to move it to the right of its forward path (in the northern hemisphere). This force, the Coriolis force, has the same practical utility as centrifugal force, and is subject to the same theoretical proviso, namely, that it vanishes when rotational accelerations are properly taken into account. The Coriolis effect enters into the motion of aircraft and projectiles, and even into the motion of a freely falling particle released from a stationary position, which will strike the ground a little to the eastward of the plumb vertical through the point of release. It also explains the well-known feature of weather charts, that outside the tropics winds tend to blow along lines of equal pressure (isobars) and not across them, as they would on a non-rotating Earth.

This 'absolute' rotation of the Earth with respect to the universe at large is in practice observed with respect to the stars, and the period of rotation is the *sidereal day*. Just as the lunar day is longer than the solar day, by virtue of the Moon's orbital motion around the Earth, so the solar day is in turn longer than the sidereal day, by virtue of the Earth's orbital motion around the Sun. Having completed one rotation with respect to the stars, the Earth has to rotate a little further in order to complete a rotation with respect to the Sun. These small additional rotations accumulate to a whole rotation in the course of one complete revolution of the Earth around the Sun. Thus one year contains $365\frac{1}{4}$ solar days and $366\frac{1}{4}$ sidereal days. Astronomers find it convenient to have clocks running to sidereal time (gaining about 4 minutes per day on the normal solar time clocks) because then the clock time shows the position of the stars and not the position of the Sun. All large astronomical telescopes are fitted with a sidereal clock drive, which compensates for the Earth's rotation, and keeps the telescope steadily trained on a particular star.

Because the Earth rotates, it is customary to envisage a fixed line running through its centre, from the North Pole to the South Pole, acting in the same way as the axle of a supported model terrestrial globe. This is quite a helpful picture when we are trying to understand the broad features of the situation, such as day and night, the seasons, the tides, and the meaning of sidereal time. It would be sufficiently accurate for the purposes of astronomers on other planets observing the Earth's rotation, and it has been sufficiently accurate so far for the Earth's astronomers observing the rotation of other planets. But this picture must be modified in order to represent the more detailed knowledge of the Earth's rotation which has been obtained, particularly in the last hundred years or so.

The first point to clear up is that we have not yet made up our minds about the exact position of the Earth's centre of gravity. It must be somewhere near the centre of figure, and so the simplest assumption to make is that these two points coincide. The difficulty is that we cannot yet be sure of the position of the North and South Poles on the Earth's surface. If there are gravity anomalies in the polar regions, as there certainly are in other regions, they will have to be measured and taken into account. The one thing that we are sure of is the direction of the Earth's axis at any given time. The stars appear to move in circles because of the Earth's rotation, and a star near the celestial pole in the sky appears to move in a very small circle. We can easily locate the centre of the circle, and if we point to that centre we can be sure that we are pointing in a direction parallel to the Earth's axis at that time. The other important direction for any observer is fixed by the apparent gravity at his station, and he can make an accurate determination of his zenith, or overhead point, by using the horizontal surface of a liquid, such as some mercury in a dish.

If the axis of the Earth were a fixed axle, turning in bearings on a fixed stand, the astronomers of today would find that their polar point and zenith point at a given place were the same as those determined by astronomers of long ago. This is not so. The first departure discovered was that the polar point is moving among the stars, which indicates that the Earth's axis is slowly changing its direction in space. At first it seemed that this motion, known as *precession*, was steadily carrying the polar point around a circle among the stars, at a fixed distance from the pole of the ecliptic (the direction perpendicular to the plane of the Earth's orbit around the Sun). Later, it was discovered that the motion was not uniform, so that the polar point oscillated backwards and forwards about its mean position, and wandered from side to side, as it travelled along the precessional circle. This additional small periodic motion received the name of *nutation*. A clear illustration of rotation, revolution, precession and nutation may be given by means of the usual terrestrial globe mounted in oblique bearings on a stand. Looking down on the northern hemisphere, spin the globe counter-clockwise to show the rotation. Carry the globe in a horizontal orbit around any object representing the Sun, travelling in the same counter-clockwise direction, and keeping the axis parallel to itself, to illustrate revolution. Now slowly turn the horizontal base of the stand in a clockwise direction to show precession changing the direction of the axis in space, but maintaining the same obliquity. Finally, rattle the axle in its bearings to illustrate the small excursions of nutation.

In the above description of precession and nutation we have not considered the cause of the axis motion, which lies in the non-sphericity of the Earth. The oblate shape of the planet can be represented as a sphere carrying an extra layer of material which is thickest around the Equator. The tidal forces of the Sun and Moon act on this equatorial bulge, and try to bring it into the plane of the Earth's orbit. The rotation of the Earth causes it to act rather like a gyroscope, so that the axis responds to this attempt to change its direction by moving sideways. The tidal forces urge the polar point towards the pole of the ecliptic, and the resulting sideways motion causes it to circle around the pole of the ecliptic in the precessional circle. The tidal forces vary slightly in magnitude and direction from time to time, and the corresponding fluctuations in the axis motion constitute the nutation.

The motions of precession and nutation have been determined with fair accuracy by analysis of observations over a long period. Their values can be reliably predicted in advance, and the necessary corrections to astronomical observations are made as a matter of routine. The actual size of the motions gives information about the masses of the Sun and

Moon, and about the distribution of mass within the Earth. It is important to note that precession and nutation do not change the position of the poles on the surface of the Earth; in other words, they do not alter the latitude and longitude of fixed stations.

It was not until 1888 that astronomers first detected the existence of a completely new motion of the Earth's axis which caused a small fluctuation in the latitude of fixed stations, and was generally known as *latitude variation*. It soon became clear, from observations of latitude made in different parts of the world, that the fluctuation corresponded to a wandering of the Earth's poles over the surface, and the motion is now called *polar variation*. The wandering extends over an area about 40 feet across, and so the extent of the motion is quite small, but it has to be allowed for in accurate work, and is troublesome because it is not regular enough to be reliably predicted. A network of observatories makes continuous observations of the polar variation, and the results are collected and published as soon as they are ready so that all interested observatories can keep their corrections up to date. The technique for observing the variation consists of the accurate location of the zenith point of a fixed station, with respect to stars of accurately known position. The measurements show a variation which is partly caused by precession and nutation, and partly by polar variation. The effect of precession and nutation is known accurately, and can be removed, leaving the remainder to be identified with the polar variation. The cause of this motion of the Earth's axis appears to be the impingement of regular and irregular meteorological disturbances, mainly air movements, upon a natural tendency for the poles to wander in a circle in a period of about fourteen months. This tendency was predicted by Euler, who calculated that the period would be ten months if the Earth were rigid. The yielding of the Earth's material to the applied stresses increases the period from ten months to fourteen. Obviously the regular meteorological changes will tend to force a twelve-month period on the wandering, which would complicate the motion. In practice the irregularities in the weather render the whole effect hopelessly unpredictable.

It is rather easier to understand the effects of polar variation if the lines of latitude and longitude on a globe are thought of as a spherical cage, supported on pivots at the North and South Poles. The globe, marked with the familiar outlines of the continents and oceans, is imagined as fitting within the cage. When demonstrating rotation, revolution, precession and nutation, the globe is firmly fixed in the cage, and moves with it. In order to show the polar variation, we must loosen the globe, and rock it slightly while holding the cage still. This shows at once that half the surface of the Earth travels downwards, away from the North Pole, and the other half moves upwards, towards the North Pole. It is also clear that certain places on the globe are carried eastward or westward by the rocking of the globe, so that there is a variation of longitude as well as a variation of latitude. For places on the Equator, however, the motion, if any, is always northward or southward, and there is no longitude variation. The motion of the globe within its cage is not predictable, but the motion of the cage is amenable to calculation, and forms the basis of astronomical tables. Astronomers naturally make observations from observatories fixed to the globe, and so they must take account of the polar variation when comparing their results with the tabulated values.

FIG 4 Polar variation is a rocking of the Earth inside its fixed 'cage' of latitude and longitude lines

After making allowances for precession, nutation and polar variation, which are motions of the Earth's axis, we might hope that the rate of rotation of the planet about that axis would remain constant, and that we could calculate all the motions of stars, planets, and the Sun and Moon for past, present and future, on a uniform time scale provided by the steadily turning Earth. Unfortunately, it has been known for a long time that this cannot be done. If we try to calculate the past motion of the Sun and Moon, and to work out when and where eclipses took place, we find that our calculations do not agree with historical records of eclipses. It appears that the motion of the Sun and Moon has been gradually speeded up over the last 2,000 years or so, and this is rather puzzling until we realize that it could be just the reflection of a slowing down in the rate of rotation of the Earth. Furthermore, if we analyse the accurate observations of Sun, Moon and planets obtained in the last 300 years or so, we find that the rate of rotation sometimes increases, and sometimes decreases, in a quite irregular pattern. When it is a question of marking out a uniform time scale for studying the motions of the Solar System, the rotation of the Earth is not a good enough clock for the purpose. The problem of establishing a uniform long-term time scale remained unsolved until quite recently, as we shall see. We are now at the point where clocks can be constructed to give a high degree of uniformity over short periods, and the periods are becoming longer as technical progress is made, so that we can almost detect the increases and decreases in the rate of the Earth's rotation. In addition to these irregular changes, it is found that there is an annual fluctuation, which is shown up quite clearly by the best modern clocks.

As to the causes of these variations, it is believed that the irregular changes arise from changes in the magnetic coupling between the Earth's core and mantle, while the annual fluctuations are ascribed to the annual changes in the velocity of high-altitude winds. The long-term slowing down has been explained as the result of the braking effect of the tidal flow in shallow seas, but recently it has been suggested that the atmospheric tides may act so as to speed up the rotation, so that there is a possibility that these two opposing effects may result in the rate of rotation settling down to a steady value.

Time

The natural way of dividing time has always been by means of nights and days, which are the visible result of the rotation of the Earth with respect to the Sun. It seemed natural to proceed with the subdivision of the day by means of a sundial, and this device served for many centuries as the main instrument of timekeeping. It required no maintenance, and could not accumulate any significant error, but it could not function at night or on cloudy days. Different means of using the stars to give the time at night were developed at various times, but the major advance came with the introduction of the mechanical clock, particularly when fitted with a pendulum. It was discovered that good clocks agreed with each other rather better than they agreed with the sundial, and the difference between clock time and sundial time was tabulated under the title *equation of time*. This discrepancy represents the non-uniformity of apparent solar time as shown by the sundial, which arises from the non-uniform velocity of the Sun along the ecliptic, and the varying inclination of the Sun's progress as it follows the inclined ecliptic. The more uniform time kept by clocks was called *mean solar time*, to indicate that it ran at the mean rate of solar time taken over a whole year (see Fig. 5, page 18).

When the running of clocks was checked by observations of stars, which were in fact much more convenient than the Sun for this purpose, the time determined was apparent sidereal time. In the twentieth century, the development of free-pendulum clocks of high precision made it necessary to allow for nutation, and the time determined was then termed *mean sidereal time*, by analogy with the mean solar time obtained by correcting apparent solar time for the Equation of Time. The effect of nutation on timekeeping is now known as the *equation of the equinoxes*, thus emphasizing the analogy still more strongly. The next step was the introduction of the quartz crystal oscillator clock, which ran even more uniformly than the pendulum clock, and therefore revealed the need to make more refined corrections to the astronomically observed time. For example, the variation of longitude caused by the polar variation appears as a variation in the astronomical time observed at a fixed observatory, and the size of this variation, though negligible when compared with the intrinsic variations found in pendulum clocks, is not negligible when compared with the more uniform running of a quartz clock. Similarly, the annual fluctuation in the rate of rotation of the Earth could hardly be detected by comparison with

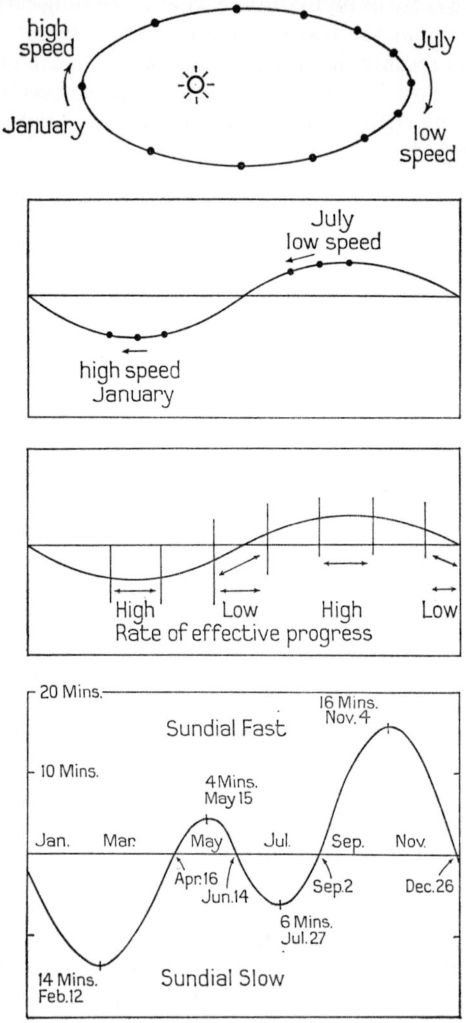

FIG 5 The equation of time consists of an annual wave plus a six-monthly wave

pendulum clocks, but is too large to be ignored in the assessment of quartz clock performance.

The latest development in timekeeping technique, the atomic frequency standard, makes the inherent precision of atomic frequencies available for giving spot checks, or a continuous check, on the running of a quartz crystal clock. Where atomic frequency standards have been designed for continuous running, they provide an alternative to the quartz crystal clock; but even if they are designed for spot-check work, rather than for continuous operation, they represent an alternative to the astronomical time observations, providing a natural standard of frequency independent of the rotation of the Earth. The modern combination of a quartz clock and an atomic resonator has given us, for the first time, a chance of detecting the irregular changes in the Earth's rotation as they happen, instead of having to wait several years for the information to emerge from observations of motions in the Solar System. There is already a tendency to relegate the rotation of the Earth to a secondary position as a timekeeping standard, and to calibrate the naturally constant atomic frequency directly against the naturally constant planetary motions.

In 1957, the basis of the legal international definition of the length of the second was transferred from the rotation of the Earth to the revolution of the Earth in its orbit around the Sun. The time scale constructed from this type of definition is called *ephemeris time*; it was introduced in 1960 as a uniform time scale in which to compute the ephemeris or time-table of planetary motions. In practice, ephemeris time is determined from the observed position of the Moon in its orbit around the Earth, as compared with the standard time-table for the Moon. Although this is the most precise method available at present, it is not capable of giving very high accuracy, and there are possibilities of doing better by using observations of artificial satellites of the Earth.

Within a few years of 1960, commercial atomic frequency standards of high reliability became available. There were several precise determinations of atomic frequencies in terms of ephemeris time, and in 1967 the definition of the second was again changed, to a form based on a specified atomic frequency. Consequently, the physicist is now able to provide himself with a 'standard second' in his laboratory, and no longer needs to rely on the astronomer to extract an approximately uniform time from the essentially non-uniform rotation of the Earth, or from the inconveniently slow planetary motions.

1.3 The Moving Planet

The distinguishing characteristic of a planet is that it moves in an orbit around a star, and the Earth is a planet because the Sun is a star, and because the Earth revolves in an orbit around it. This, however, is a relatively modern view of the situation, dating from the sixteenth century. In earlier times it was taken as obvious that the Sun was different from the stars and the planets from the Earth, and that the whole heavens, with Sun, Moon, planets and stars moved around the fixed central Earth. The five planets known in classical times were Mercury, Venus, Mars, Jupiter and Saturn. It is true to say that the next planet to be discovered was the Earth itself.

When we were considering the rotation of the Earth on its axis, we found that there were different types of day with various lengths, such as the solar day, the lunar day and this sidereal day, depending on the reference point used to define the completion of one rotation. Similarly, when considering the orbital motion of the Earth around the Sun, there is a choice of various reference points, and a corresponding choice of different types of year with various lengths. The most conspicuous indicator of the passage of a year is the cycle of the astronomical seasons, so that we may take this period, known as the *tropical year*, as the most familiar type of year. During the course of a tropical year, each of the Earth's two poles is presented in turn to the Sun; in other words, the Sun approaches and recedes from the north pole, and then approaches and recedes from the south pole. When the Sun is just halfway between the poles it is overhead at the Equator, giving equal days and nights all over the Earth. This state of affairs, called an *equinox*, occurs twice in each tropical year; there is the *vernal equinox* in March, when the Sun is crossing the Equator from south to north, and the autumnal equinox in September, when it crosses from north to south. The closest approach of the Sun to the north pole marks the *summer solstice* in June, and the closest approach to the south pole the *winter solstice* in December. At these times days or nights are at their longest in one hemisphere and shortest in the other, and the Sun is overhead at one or other of the tropics. The astronomical seasons are the intervals between the equinoxes and the solstices and they have no precise connection with weather.

The length of the tropical year is 365·2422 mean solar days, and the problem immediately arises of arranging the calendar, which can only count whole days, so that it remains in step with the cycle of the seasons. If we make each calendar year count 365 days, neglecting the odd fraction, it will not take long for the calendar to run out of line with the seasons, which would appear to arrive a week later in the calendar every 29 years. This was the trouble that Julius Caesar tried to cure by introducing a leap year of 366

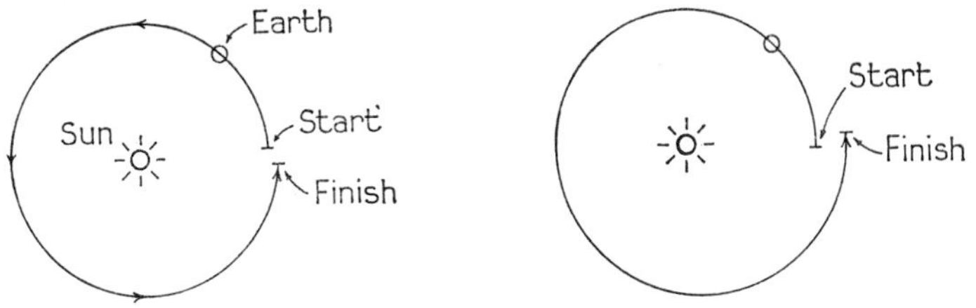

Common year (365 days) Leap year (366 days)

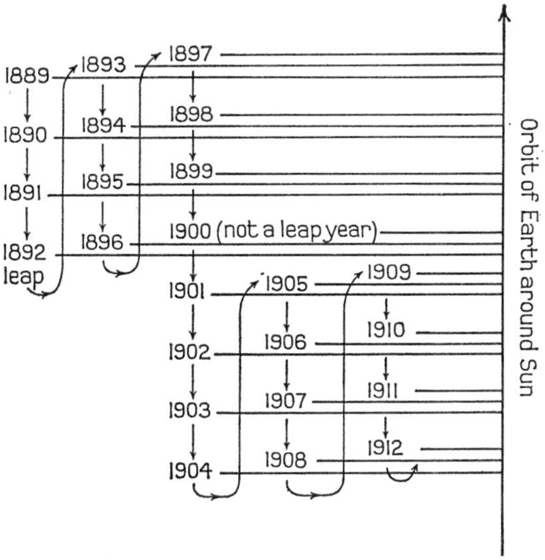

FIG 6 Keeping the calendar right. Starting points of calendar years 1889–1912

days after each three ordinary years of 365 days. The calendar now counted 1,461 days as four years, giving an average length of 365·25 days per year, and this system was good enough to last for 16 centuries. The correction made was a little too large, so that the seasons gradually arrived earlier in the calendar, at the rate of about three-quarters of a day in each century. Eventually the accumulated error began to make it awkward to decide when Easter should fall, and Pope Gregory XIII adjusted the calendar by omitting ten days from the calendar in October, 1582, and by arranging to cancel three leap years in every four centuries. The years 1600 and 2000 were retained as leap years, but 1700, 1800 and 1900 were made ordinary years. This Gregorian calendar counts 146 097 days as four hundred years making the average length of the year work out at 365·2425 days. The seasons still gradually arrive earlier in the calendar, but at the rate of only one day in about 3,000 years. The change to the Gregorian calendar was not made in Great Britain until 1752, by which time the discrepancy had increased to eleven days. The omission of these days from the month of September, 1752, caused a great deal of controversy and some civil commotion. The Eastern Orthodox churches still observe their festivals by the unadjusted Julian calendar.

Owing to the precessional movement of the Earth's axis, the tropical year is slightly shorter than the absolute period of revolution around the Sun, the *sidereal year*, which is 365·2564 mean solar days. The reference point is this case is a fixed star, whereas the tropical year is referred to the equinoxes and solstices, which have a precessional movement relative to the fixed stars. The small difference in the lengths of the two types of year was observed by Hipparchus, and correctly explained by him as arising from precessional motion. As previously described, the effect of the precessional motion is to carry the polar point in a circle around the pole of the ecliptic, and one circuit will take about 26 000 years. It follows that 13 000 years ago, and again in 13 000 years from now, the north celestial pole was, and will be, near the bright star Vega, about 47 degrees from the present position of the pole. The equinoxes will have changed places in the sky, and so will the solstices. The stars which are now high in the sky on winter nights will be low in the sky on summer nights, and so on. Orion will have almost sunk from view for observers in the British Isles, but the Southern Cross will have come north to take his place.

The third type of year is the *anomalistic year*, the period of revolution with respect to the perihelion of the Earth's orbit, which is the point of the orbit nearest to the Sun. The orbit is an ellipse, with the Sun at one focus, so that the nearest and furthest points of the orbit, the perihelion and the aphelion, lie at opposite ends of the major axis, which is the longest diameter of the orbit. If the ellipse remained stationary in space, relative to the stars, the anomalistic year would be the same length as the sidereal year. Owing to the gravitational action of the other planets upon the Earth, however, the ellipse is slowly turning in the same direction as the revolution of the Earth, so that it takes a little longer for the Earth to catch up with the perihelion point. The length of the anomalistic year is 365·2596 mean solar days.

The fourth type of year, the *eclipse year*, has a length of only 346·6200 mean solar days, and is the period of revolution of the Earth with respect to the nodal line of the Moon's orbit. This nodal line is the intersection of the plane of the Moon's orbit around the Earth with the plane of the Earth's orbit around the Sun. Eclipses of the Sun and Moon can only occur near the times when the nodal line passes through the Sun, at intervals of half an eclipse year. The extreme shortness of the eclipse year indicates that the nodal line is rotating fairly rapidly in the opposite direction to the Earth's revolution, and in fact it makes a complete rotation of 360° in about 19 years.

Having traced the motion of the planet Earth on its axis, and its revolution around the Sun, the next step would be to examine the motion of the Solar System in the galaxy and finally the motion of the galaxy itself. But this is beyond the purpose of this book. It seems certain that our galaxy is in some sort of motion through space, but the only objects by which we may judge or measure this motion are the other galaxies. There is some evidence that galaxies are clustered together in groups, but the individual galaxies in each group have individual notions within the groups, and if we try to compare the motion of one group with another, we come up against the fundamental difficulty that all such groups of galaxies appear to be receding from one another. It may be that they are moving apart through space, but the modern interpretation is that, in some almost inconceivable way, they are being carried apart by the expansion of space itself. At this point the whole idea of motion has to be reconsidered in the light of cosmology and the theory of relativity, and so it is a very convenient point to call a halt, and to return from the depths of space to the Earth itself.

Geodesy

A. H. Cook, Professor of Geophysics,
University of Edinburgh

2.1 Introduction

Geodesy is the study of the external shape of the Earth as a whole. It provides the geometrical framework into which other knowledge of the Earth may be fitted and, in its own right, yields information about the internal constitution of the Earth. It might seem that there can be no doubt about what is meant by the shape of the Earth, but of the entire surface, no more than about one quarter is land and the methods of ordinary geometrical survey cannot be applied to the shifting surface of the seas. Yet that surface is held in its place by the force of gravity and the shape of the Earth over the greater part of its area means the shape of a surface controlled by the gravitational attraction of the Earth. Here lie the connections between geodesy and celestrial mechanics on the one hand and the internal structure of the Earth on the other.

Were the Earth bounded by a solid surface almost everywhere, it might be supposed that it could be surveyed by selecting points upon it to form the apices of a polyhedron, the angles and sides of which could then be measured. This was indeed the programme of early geodesy, during the first part of the eighteenth century, for example, but it was found to be impracticable. The curvature of the Earth and the optical scattering of the atmosphere limit direct measurements between points on the surface of the Earth to rays of some 50 kilometres in length; optical measurements of angles and distances are accordingly affected by atmospheric refraction. Consider measurements of angles and distances in a triangle formed by three points of the imaginary polyhedron (Figure 7). The points lie nearly in a horizontal plane, and the rays are therefore almost tangential to the surfaces of constant air density which run nearly parallel to the surface of the Earth. All optical rays between the vertices are therefore slightly curved in a vertical plane, the lengths, if measured by optical transit time methods, need small corrections, the horizontal components of the angles are almost unaffected by refraction, but the vertical components of the angles are seriously affected by refraction. On account of atmospheric refraction alone, purely geometrical measurements close to the surface of the Earth cannot be expected to give a reliable survey of the shape of the solid surface. The course which has been followed since all but the earliest essays in geodesy, has been to make the local geometrical measurements with respect to a reference surface that can be determined on a world-wide basis. The purely geometrical measurements need then have solely local validity and errors in them do not vitiate the determination of the over-all shape of the Earth. The world-wide reference surface is related to the gravitational field of the Earth.

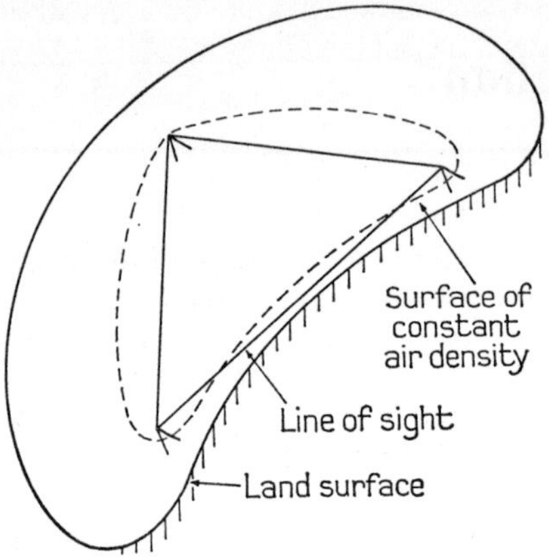

FIG 7 Survey points, lines of sight and surfaces of constant air density

The gravitational attraction of the Earth can be derived from a potential, that is to say, if g_x, g_y, g_z are the components of gravity in a Cartesian coordinate system, they may be written as

$$g_x = -\partial V/\partial x$$
$$g_y = -\partial V/\partial y$$
$$g_z = -\partial V/\partial z$$

where V is the potential. In vector notation,

$$\mathbf{g} = -\text{grad } V \quad \text{or} \quad -\nabla V.$$

The observed gravitational acceleration at the surface of the Earth is not, however, just that due to the Newtonian attraction, for the Earth is spinning about its polar axis with an angular velocity ω and therefore a body at the surface has an acceleration at right angles to the polar axis; in terms of polar coordinates (r, θ, λ) where r is the radius vector from the centre of the Earth, θ is the co-latitude and λ is the longitude, the components of the acceleration are

$$(r\omega^2 \sin^2 \theta, r\omega^2 \cos \theta \sin \theta, 0).$$

These also can be derived from a potential, for

$$r\omega^2 \sin^2 \theta = -\partial\{-\tfrac{1}{2}r^2\omega^2 \sin^2 \theta\}/\partial r,$$

$$r\omega^2 \sin \theta \cos \theta = -\frac{1}{r}\,\partial\{-\tfrac{1}{2}r^2\omega^2 \sin^2 \theta\}/\partial \theta.$$

Thus the resultant acceleration at the surface of the Earth, due to the Newtonian attraction and the rotational acceleration, is given by

$$\mathbf{g} = -\text{grad}\,(V - \tfrac{1}{2}r^2\omega^2 \sin^2 \theta).$$

Surfaces on which the combined potential (known as the *geopotential*) takes a constant value are called equipotential surfaces. The surface of a liquid at rest is a surface of constant potential, for if it were not, work would be done by the forces derived from the potential and the liquid would flow until its surface was one of constant potential. The surface of the sea is not one of constant potential because it is disturbed by waves, tides and currents, but if the effects of these can be averaged out over a long time, we may define an equipotential surface, the mean sea level surface. To this surface known as the *geoid*, geodetic measurements are referred.

The sea level surface is however very far from ideal. In the first place, it is doubtful how well it can be defined. Wave disturbances can be averaged out given observations over a sufficiently long period. The forces causing tides are indeed derived from a potential, that of the varying attractions of the Sun and the Moon at the surface of the Earth, a potential that is not a constant but varies periodically with time. Over the deep oceans the surface of the sea follows this varying potential but, close to land, while the variations are still periodic, the effects of the shores on the movements of the water distort and amplify the changes in the height of the surface. Lastly, systematic currents, driven by combinations of forces due to winds, the rotation of the Earth and differences of density within the oceans, cause the surface to depart systematically from one of constant potential. Thus, especially close to land where for survey purposes, the information is most needed, it is difficult to derive from the actual surface of the sea as recorded by a tide gauge for example, a reliable estimate of the position of the surface of constant geopotential coincident with the mean surface of the sea. But there is a further difficulty. Only close to the shores can the geoid be thus located and the main land areas which are subject to geodetic survey are beyond the reach of this realization. Fortunately, as will be seen, knowledge of the position of the geoid can be carried into land areas by making use of the fact that the free surface of the liquid in the bubble of a spirit level lies in an equipotential surface of the geopotential, though in general one on which the value of the geopotential is different from that on the geoid. The spirit level is also not subject to disturbances of waves and currents. The geoid itself is not accessible on land areas for in general it will lie below the surface within the terrestrial matter, but with the help of the spirit level, the positions of other equipotential surfaces outside the solid matter can be located and local geometrical surveys can be referred to them. With the aid of a theory of the external gravity field of the Earth and of observations of gravity, the equipotential surfaces passing through one area of local survey may be related to the geoid and to those passing through other local areas, and geometrical surveys may be put into proper relationship to each other. The questions to be discussed in this chapter are therefore the ways in which local surveys are related to local equipotential surfaces, and the theory and practice of the determination of the equipotential surfaces from gravity measurements. The second question is the more difficult for the theory is not straightforward; at the same time, the knowledge of the gravity field of the Earth that is obtained tells us something of the interior state of the Earth. Lastly the problem of a geometrical polyhedral framework for the Earth will recur, for it is now possible to go a long way to realizing it by means of observations of artificial satellites.

Local geometrical surveys depend on measurements made in planes tangent to local equipotential surfaces. The basic geometry is shown in Figure 8. At some point P of the survey network, the tangent plane to the

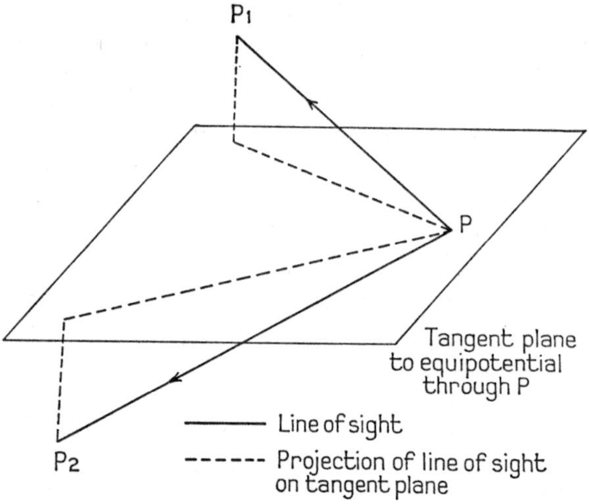

FIG 8 Geometry of local surveys

equipotential surface passing through that point is located by observations on the spirit level of a theodolite set up there. The telescope of the theodolite is pointed at other stations of the survey and the angles that are measured are those between the projections onto the tangent plane through P of the directions from it to the other stations. The difference of potential between one point of a survey and another is found as shown in Figure 9. E_1 and E_2 are the equipotential surfaces passing through the points P_1 and P_2. At P_1 a graduated staff is set up and is viewed through a telescope with its line of sight tangential to E_2 at P_2. If the points are close enough for the curvature of the equipotential surfaces to be ignored, the difference between the potentials at P_1 and P_2 is $\int_0^H g.dh$ where g is the value of gravity at

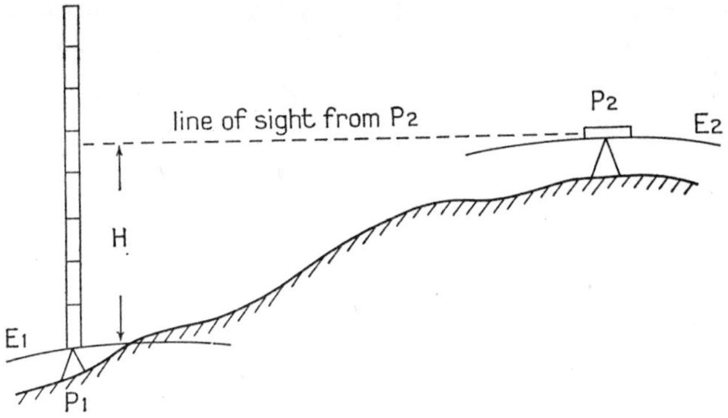

FIG 9 Determination of potential difference

height h above P_1 and H is the height above P_1 at which the line of sight from P_2 intersects the staff. The local values of gravity may readily be measured and the differences of potential accordingly derived. This type of measurement is called *geodetic levelling* and it is the basis of the heights to be found on Ordnance Survey Maps for example. Fundamental measurements are carried along lines between markers, known as *bench marks* firmly implanted in stable ground from which local subsidiary measurements originate. The fundamental lines start from tide gauges on the coast so that the values of the potential may be related to the position of mean sea level and the zero for heights in this country is located in the tide gauge at Newlyn. The heights given on maps are really potential differences and are expressed (as *orthometric* heights) in feet or metres by using a conventional formula for the variation of gravity and potential outside the geoid.

The aim of local geometrical surveys is to map points on the surface of the Earth on an ideal simple surface close to the geoid. The geoid is not convenient because it has no simple shape but it is very close to a spheroid of revolution and the actual physical points of the Earth's surface are mapped on to a spheroid adopted by convention so that it is close to the geoid. To a good approximation, the differences between the geoid and the conventional spheroid may often be ignored. The surveyor has thus to project angles measured in horizontal tangent planes down onto the geoid and he has to project distances measured between points such as P_1 and P_2 onto the geoid. Until very recently, the measurement of distances between points separated by say 30 km was a matter of the greatest difficulty and took a very great deal of time so that the positions of points in plan were determined by angular measurements in triangles linking the points and only one or two sides in the whole network were measured to give a scale to the system. These few special measured distances were known as *base lines* and the actual measured length at some height above the geoid was reduced to the geoid by projecting along normals to the conventional close fitting spheroid. The build up of errors in large sets of triangles in which only angles are measured, means that while a base line of 30 km can with great care be measured with a relative error of little more than one part in a million, the errors at some distance from it are much greater. Modern instrumental developments now permit the rapid measurement of such distances directly by observations of the travel times of electromagnetic radiation, light or radio waves, and while the accuracy of an individual line may not be as good as a special base line measurement, the overall consistency of scale in a set of triangles most of the sides of which are measured, is vastly improved.

While as will be seen in more detail below, the form of the geoid can be found from the variation of gravity over it, it may also be found by direct geometrical methods. In Figure 10, *PN* is the normal to the equipotential

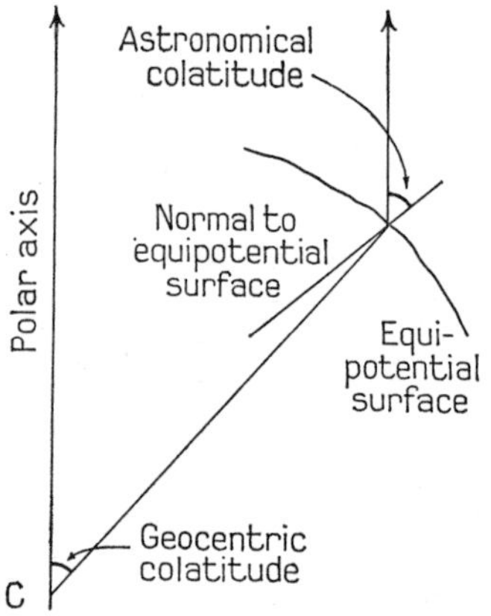

FIG 10 Determination of astronomical coordinates

surface through *P*; by definition, it is the direction of the vertical at *P*, and is perpendicular to the plane defined by the spirit level of an instrument set up at *P*. The direction of *PN* can be found also in relation to the axis of spin of the Earth by reference to the stars; the colatitude, or 90° minus the *latitude* of *P*, is the angle between *PN* and the polar axis, that is between *PN* and the star *Polaris* to a first approximation. It must be noted that the astronomical latitude and longitude that define the direction of the normal to the equipotential surface are not the same angles as the geocentric latitude and longitude that define the direction of the radius vector to *P* from *C*, the centre of the Earth. The two sets of angles would coincide only if the Earth were truly spherical. Given the direction of the equipotential surface at points not too far apart and the distances between such points over the surface, it is possible to determine the form of the geoid and that was in fact the method employed until the knowledge of gravity on the surface of the Earth had advanced so such an extent that it was possible to make useful deductions of the shape of the geoid from variations of gravity.

Until relatively recently, the methods of geodesy were almost entirely geometrical. By measurements of angles of triangles in a network covering the land, combined with a few measurements of the lengths of selected sides, the positions of points in plan were obtained, while the differences of geopotential were derived from spirit levelling. The form of the geoid, in practice really only the ellipticity of meridional sections, was obtained from observations of the direction of the local vertical relative to the stars, combined with arc lengths over the surface found from triangulation. The principal radii of curvature of the equipotential surfaces so found led to values for the polar and equatorial radii of the Earth. The principles of geodetic survey on these lines were already established at the end of the eighteenth century, at which time they were used in the surveys in France and Spain on which the definition of the Metre was based (*Base du Système Métrique*), and geodetic survey has continued on the same lines until the very recent past. While those methods are broadly speaking adequate, though in some respects laborious, for the production of maps of land areas, they suffer from one grave defect for any world-wide scientific application: the areas in which they can be applied are relatively small and are disconnected, so that the surface of the geoid is poorly sampled and reliable values for the parameters of the Earth as a whole are not to be obtained. Emphasis is placed in this chapter on the modern developments in geodesy which have given far more reliable world-wide data and also have enabled local measurements to be made much more expeditiously, that is to say, the development of methods for the measurement of distance based on the travel time of electromagnetic waves, the theory of the shape of external equipotential surfaces, modern methods of gravity measurement and the derivation of the gravitational potential of the Earth from the orbits of artificial satellites.

2.2 Theory of the External Gravity Field of the Earth

It is obvious that if the equipotential surfaces were specified geometrically in some coordinate system, the potential at any point of space could be regarded as a known function of position; the magnitude and direction of gravity would also be known since the gravity vector is normal to the equipotential surface at a point and the magnitude of gravity is inversely proportional to the separation of the equipotential surfaces. Similarly, if the potential were known as a function of position, the shape of a surface of constant potential could be obtained and the gravity vector could be calculated for any point in space. It is less obvious that if the value of gravity is known over a surface of constant potential, then the form of that surface is implied and the variation of gravity and potential outside it may be derived; that that is so will now be shown.

Let positions be specified in a system of spherical polar coordinates with the origin at the centre of the Earth. r is the radius, θ is the geocentric colatitude measured from the North Pole and λ is the longitude measured from the direction of the meridian of Greenwich. Let V be the potential of the Newtonian attraction of the matter of the Earth and let U be the geopotential so that

$$U = V - \tfrac{1}{2}r^2\omega^2 \sin^2 \theta.$$

Consider first a simplified model Earth in which the material boundary coincides with an equipotential surface so that all equipotential surfaces outside that one lie in space free from matter, in which therefore, the potential satisfies Laplace's equation:

$$\frac{\partial^2 V}{\partial r^2} + \frac{1}{r^2}\frac{\partial^2 V}{\partial \theta^2} + \frac{1}{r^2 \sin^2 \theta}\frac{\partial^2 V}{\partial \lambda^2} = 0.$$

The solutions of Laplace's equation expressed in spherical polar coordinates are spherical harmonics:

$$V = r^n\, S_{nm}(\theta, \lambda) \text{ or } r^{-n-1} S_{nm}(\theta, \lambda),$$

where S_{nm} is a surface harmonic:

$$S_{nm}(\theta, \lambda) = P_n^m(\cos \theta)(\sin m\lambda, \cos m\lambda)$$

and $P_n^m(\cos \theta)$ is an Associated Legendre Function.

Since the potential of an isolated body like the Earth must die away to zero at very great distances, the potential V outside the Earth may be written as the sum:

$$V = \sum_{n} r^{-n-1} \sum_{m} a_{nm} S_{nm}(\theta, \lambda).$$

The potential of the rotational acceleration may be written as a spherical harmonic;

$$\tfrac{1}{2} r^2 \omega^2 \sin^2 \theta = \tfrac{1}{3} r^2 \omega^2 \{1 - P_2(\cos \theta)\}.$$

Thus the geopotential also satisfies Laplace's equation in the free space outside the matter of the Earth.

Since the Earth departs from a sphere by no more than about 1 in 300, the radius vector of an equipotential surface is nearly constant and may be written as

$$r = r_0\left(1 + \sum_{n,m} \rho_{nm} S_{nm}(\theta, \lambda)\right)$$

where r_0 is the mean radius of the equipotential surface. The condition that a surface should be an equipotential is simply that if the value for r on such a surface is put into the general expression for the potential, the result should be a constant. That condition will now be written down for the particular case of the geoid, the sea level equipotential surface which for the present model is assumed to contain all the matter of the Earth. Then

$$\sum_{n=0}^{\infty} \left\{ r_0 \left[1 + \sum_{p,q} \rho_{pq} S_{pq}(\theta, \lambda) \right] \right\}^{-n-1} \sum_{m} a_{nm} S_{nm}(\theta, \lambda)$$
$$- \tfrac{1}{3} \omega^2 r_0^2 \left[1 + \sum_{p,q} \rho_{pq} S_{pq}(\theta, \lambda) \right]^2 \{1 - P_2(\cos \theta)\} = C.$$

This rather formidable expression may be simplified considerably. In the first place, the potential V at a great distance from the Earth is just $-GM/r$ (G is the Newtonian constant, M the mass of the Earth) so that

$$a_{00} = -GM.$$

Secondly, since in practice all the departures from spherical symmetry are small, the radius r may be replaced by r_0 in all except this principal term in the potential. Accordingly,

$$-\frac{GM}{r_0}\left[1 - \sum_{n,m} \rho_{nm} S_{nm}\right] + \sum_{n,m} \frac{a_{nm}}{r_0^{n+1}} S_{nm} - \tfrac{1}{3} \omega^2 r_0^2 [1 - P_2(\cos \theta)] = C.$$

It follows that $C = -GM/r_0 - \tfrac{1}{3}\omega^2 r_0{}^2$,

$$\rho_2 \frac{GM}{r_0} + \frac{a_2}{r_0^3} + \tfrac{1}{3} \omega^2 r_0^2 = 0,$$

and for all other harmonic terms

$$\rho_{nm}\frac{GM}{r_0} + \frac{a_{nm}}{r_0^{n+1}} = 0.$$

These are the simple relations which relate the form of the equipotential surface to the variation of the potential in the special case of surfaces which depart only slightly from spheres. The results have been taken only to the first order in the departures from a sphere. That is adequate for all except the term proportional to $P_2(\cos\theta)$, for those terms are of order 10^{-6}, but it is necessary to go to a second approximation for the term in $P_2(\cos\theta)$, which is the one that determines the ellipticity of a meridian and is of order 10^{-3}.

The notation can be simplified by writing the potential as

$$-\frac{GM}{r}\left[1 - \sum_n \left(\frac{r_0}{r}\right)^n \sum_m J_{nm} P_n^m(\cos\theta) \cos \mathbf{m}(\lambda - \lambda_{nm})\right]$$

where

$$\frac{\omega^2 r_0^3}{GM} = \mathbf{m}.$$

Then

$$\rho_{nm} = J_{nm},$$

$$\rho_2 = -J_2 - \tfrac{1}{3}\mathbf{m}.$$

The radius vector on an ellipsoid of revolution is given by

$$r = a_e(1 - \tfrac{1}{3}f)[1 - \tfrac{2}{3}fP_2(\cos\theta)],$$

where a_e is the equational radius and f is the flattening, equal to $(a_e - b)/a_e$; $r_0 = a_e(1 - \tfrac{1}{3}f)$.

Hence

$$\rho_2 = -\tfrac{2}{3}f$$

and

$$f = \tfrac{3}{2}J_2 + \tfrac{1}{2}\mathbf{m}$$

or

$$J_2 = \tfrac{2}{3}f - \tfrac{1}{3}\mathbf{m}.$$

The value of gravity, if only terms of the first order are to be retained, is given by

$$-\frac{\partial U}{\partial r} = -\frac{GM}{r^2} + \frac{3GM}{r^2}\left(\frac{r_0}{r}\right)^2 J_2 . P_2(\cos\theta) + \tfrac{2}{3}\omega^2 r[1 - P_2(\cos\theta)].$$

and thus on an equipotential surface it is

$$-\frac{GM}{r_0^2}\left[1 - 2\rho_2 P_2(\cos\theta)\right] + \frac{3GM}{r_0^2} J_2 P_2(\cos\theta) + \tfrac{2}{3}\mathbf{m}\frac{GM}{r_0^2}\left[1 - P_2(\cos\theta)\right]$$

$$= -\frac{GM}{r_0^2}[1 - \tfrac{2}{3}\mathbf{m} - \{-\tfrac{4}{3}f + 3J_2 - \tfrac{2}{3}\mathbf{m}\}P_2(\cos\theta)].$$

Thus

$$g = \gamma_0\{1 + (\tfrac{5}{3}\mathbf{m} - \tfrac{2}{3}f)P_2(\cos\theta)\}$$

or

$$g = \gamma_0\{1 + (\tfrac{4}{3}\mathbf{m} - J_2)P_2(\cos\theta)\}$$

where

$$\gamma_0 = \frac{GM}{r_0^2}(1 - \tfrac{2}{3}\mathbf{m}) = \frac{GM}{a_e^2}(1 - \tfrac{2}{3}\mathbf{m} + \tfrac{2}{3}f)$$

$$= \frac{GM}{a_e^2}(1 + J_2 - \tfrac{1}{3}\mathbf{m}).$$

Alternatively,

$$g = \gamma\{1 + (\tfrac{5}{2}\mathbf{m} - f)\cos^2\theta\}$$
$$= \gamma\{1 + (2\mathbf{m} - \tfrac{3}{2}J_2)\cos^2\theta\}$$

where

$$\gamma = \frac{GM}{a_e^2}(1 - \tfrac{3}{2}\mathbf{m} + \tfrac{4}{3}f)$$

$$= \frac{GM}{a_e^2}(1 + 2J_2 - \tfrac{5}{6}\mathbf{m}).$$

If all the other terms in the potential are included, the value of gravity at the surface is found to be

$$\gamma\{1 + (2\mathbf{m} - \tfrac{3}{2}J_2)\cos^2\theta\} + \sum_{n,m}(n + 1)\frac{GM}{a_e^2}J_{nm}P_n^m(\cos\theta)\cos m(\lambda - \lambda_{mn}).$$

From these formulae it will be seen that the form of the surface may be found from the variation of gravity over it or from the variation of the potential outside it, or that the potential may be calculated from the shape of the surface or from the variation of gravity upon it. The formulae as just given for the harmonic term $P_2(\cos\theta)$ are not accurate enough for practical purposes and it is necessary to work them out to include the terms with J_2^2 or f^2; the calculations just given shew perfectly well, however, the physical principles of the calculations.

The foregoing results are a first approximation. They apply to the sea level surface where it is accessible but they do not apply to the geoid where it goes beneath the land because it was assumed in the calculations that no matter lay outside the surface of which the form was being found. Now the forms of the equipotential surfaces that lie outside the topographical surface of the solid Earth are those that are really required, that is the forms of those equipotential surfaces of the gravity field that do lie outside the matter of the Earth and that do satisfy Laplace's equation; and it can be shown that the properties of the external field may be calculated from a model in which it is supposed that the geoid does include all the matter of the Earth, none lying outside it, and that the values of gravity on it are not the actual ones, which in any case are not known, but those on the actual surface increased by the amount $\delta V/a_e$, where δV is the difference between the potential on the geoid and that at the external point. The quantity $\delta V/a_e$ is known as the *free air correction*, values of gravity to which it is added are known as *free air values of gravity* and differences of such values from some conventional value of gravity are known as *free air anomalies*. In the practical applications of the result, it is usually the form of the geoid that is calculated from the free air values of gravity although it will be appreciated that it is the external surfaces just above the topographic surface to which the result applies. For the most part, the errors will not be great.

Mathematical methods have been developed, on the basis of the theory outlined above, for the calculation of the shape of the geoid from the variation of gravity over it. The calculations could be done by expressing the anomalies as a sum of harmonic terms, working out each corresponding term in the radius vector of the geoid, and finally adding together all the harmonic terms in the radius vector. However, values of gravity are not known over much of the surface of the Earth and so the harmonic terms in the series for the gravity anomalies cannot be found very reliably, and other ways of treating the data are available. The result is that it is possible to calculate from gravity anomalies the shape of the geoid in rather restricted areas over and near which the values of gravity have been well surveyed, but it is not possible from gravity data to obtain a good overall picture of the shape of the geoid. On the other hand, the external potential of the Earth can be derived from the behaviour of artificial satellites and then the overall shape of the geoid can be calculated from those results.

The determination of the potential and of surface gravity

It has been seen that geometrical methods enable the form of the external equipotential surfaces to be found and that gravity can be measured on the surface but that the results of both these procedures were not very reliable when they were applied to the Earth as a whole. The advent of artificial satellites revolutionized much of geodesy because it provided a means by which the external potential could be found conveniently with high accuracy on a world-wide scale, so that it has become possible to obtain reliable ideas of the large scale features of the geoid; in particular, the value of the polar flattening of the Earth derived from satellite data is significantly different from, and far more accurate than, the value formerly accepted on the basis of geometrical and surface gravity measurements.

Artificial satellites move subject to the gravitational attraction of the Earth and in the absence of any other forces, the acceleration of the satellite would be equal to the gradient of the potential:

$$\ddot{r} = -\text{grad } V.$$

(The spin angular velocity of the Earth does not enter because the satellite is moving freely and is not rigidly attached to the surface of the Earth). It might be thought that the potential could be found by calculating the acceleration of the satellite from observations of its position and then deriving the potential from the components of the acceleration. That would be impracticable. In the first place, the gravitational attraction is not the only force acting on the satellite (for example it is also subject to air drag), and it would be difficult to separate the gravitational force from others without taking account of special features of the various forces. Secondly, the gravitational potential of the Earth is very nearly $-GM/r$, and the parts which vary with latitude and longitude, which are the ones which correspond to departures of the surface of the Earth from the spherical form, are not more than one part in one thousand of the $-GM/r$ part, and so would be quite difficult to determine by direct observation. However, it is well known that the path of a satellite in a potential $-GM/r$, is an ellipse. To a first approximation, the potential of the Earth is the same as that due a point with a mass equal to that of the Earth located at the centre of the Earth, and to the same order of approximation, the orbits of artificial satellites are Keplerian ellipses with one focus at the centre of the Earth. Because the potential of the Earth departs slightly from $-GM/r$ and because the satellite is acted on by the drag of the atmosphere and by other non-gravitational forces, the orbits show slight differences from ellipses but the types of departure are not the same for the gravitational and drag forces so that it is quite easy to calculate the gravitational potential free of errors arising from air drag. That is fortunate since the air drag forces are relatively large, very variable and not well determined in detail.

A diagram of a satellite orbit is shown in Figure 11. An elliptical orbit is specified by its size, its ellipticity and by the time taken by a satellite to make one revolution, as well as by the position of the satellite at the instant from which time is measured. In addition the position of the orbit in space must be specified, usually by the position of one focus (the centre of the Earth), the line in which the orbit cuts the plane of the equator, the angle between the equator and the orbit, and lastly, the direction of the major axis of the ellipse. The parameters by which the orbit is defined are known as the *elements* of the ellipse; they are

- a_s the semi-major axis,
- e the eccentricity, equal to $(a_s^2 - b_s^2)/a_s^2$,
- n the mean angular velocity of the satellite in its orbit (mean motion).
- i the inclination of the plane of the orbit to the equator,
- Ω the longitude of the ascending node, measured along the equator from some direction fixed in space,
- ω the longitude of perigee measured along the orbit from the ascending node,
- τ the time at which the satellite passes through perigee.

The elements of a Keplerian ellipse described in a potential $-GM/r$ are all constant, the shape of the orbit, its size and its orientation in space all remaining fixed. The effect of a departure from the simple potential is that the elements of the orbit change with time, mostly in a periodic manner, but there may also be some steady changes. The dominant effects of air drag are steady changes in the semi-major axis, eccentricity and mean

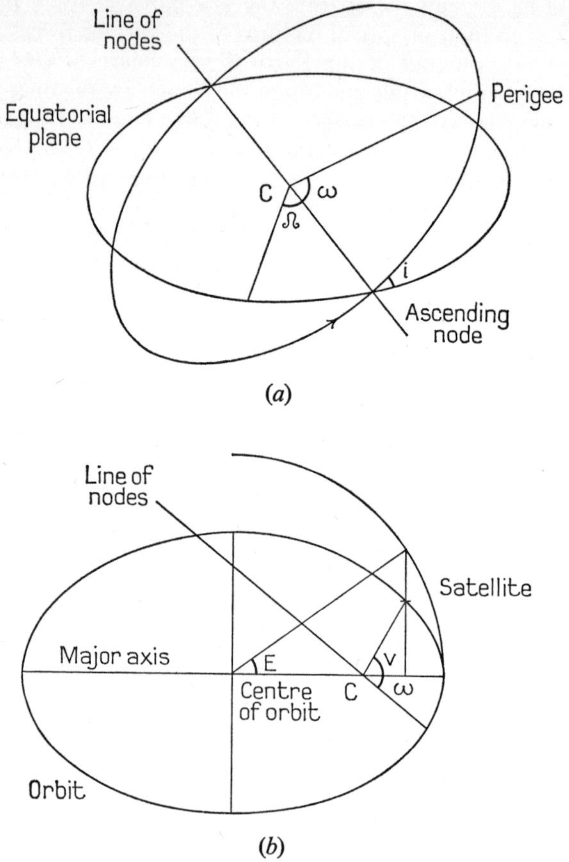

FIG 11 Geometry of elliptical orbit
(a) The orbit in space
(b) Positions in the orbit

motion, while the dominant effects of the gravitational field of the Earth are steady changes in the longitudes of the node and of perigee. It is convenient to express the forces acting on the satellite in orthogonal components, F_t tangential to the orbit, F_r normal to the tangent in the plane of the orbit, and F_n perpendicular to the other two, that is to say at right angles to the plane of the orbit. In terms of these forces, the equations of motion are

$$\ddot{r}_t = F_t$$
$$\ddot{r}_r = F_r$$
$$\ddot{r}_n = F_n,$$

and by expressing the vector $\ddot{r}$ in terms of the elliptic elements, it may be shown that the elements vary according to the following equations, known as *Lagrange's equations of planetary motion.*

$$\dot{a}_s = \frac{2}{n\eta}\{F_r e \sin v + F_t(1 + e \cos v)\},$$

$$\dot{e} = \frac{a_s\eta}{n}\{F_r \sin v + F_t(\cos v + \cos E)\},$$

$$\frac{di}{dt} = \frac{rF_n \cos u}{na_s^2\eta},$$

$$\dot{\Omega} = \frac{rF_n \sin u}{na_s^2\eta \sin i},$$

$$\dot{\omega} = 2 \sin^2 \tfrac{1}{2}i . \dot{\Omega} + \frac{\eta}{na_s e}\left[-F_n \cos v + F_t\left(1 + \frac{a_s r}{b_s^2}\right) \sin v\right],$$

v is the *true anomaly*, u the *argument of latitude*, E the *eccentric angle* (Figure 11) and $\eta^2 = 1 - e^2$.

To a first approximation, the force due to air drag is tangential to the orbit, and so the planetary equations show that air drag does not lead to any change in the longitude of the node or perigee. On the other hand, the only forces with components perpendicular to the plane of the orbit are gravitational ones and so to first order the gravitational field acting on the satellite may be obtained from observation without having to apply corrections for the effects of the atmosphere (in a higher approximation, these conclusions are slightly modified by an effect of the rotation of the atmosphere).

Let us calculate the motion of the node of a circular orbit in a potential, the only departure of which from the simple form, $-GM/r$, is the second zonal harmonic term,

$$+\frac{GM}{r}\frac{a_e^2}{r^2}J_2P_2(\cos\theta).$$

The three orthogonal components of the perturbing force in spherical polar components are

$$F_r = \frac{\partial V}{\partial r} = -\frac{3GM}{a_s^4}a_e^2J_2P_2(\cos\theta),$$

$$F_\theta = \frac{1}{r}\frac{\partial V}{\partial \theta} = -\frac{3GM}{a_s^4}a_e^2J_2 \cos\theta \sin\theta,$$

$$F_\lambda = 0.$$

Now from spherical trigonometry,

$$\cos\theta = \sin u \sin i$$

so that

$$F_r = -\frac{3GM}{a_s^4} a_e^2 J_2(\tfrac{3}{2}\sin^2 u \sin^2 i - 1)$$

$$F_\theta = -\frac{3GM}{a_s^4} a_e^2 J_2 \sin u \sin i \sin\theta.$$

Now, since $F_\lambda = 0$, $F_n = F_\theta \sin\chi$ where

$$\sin\chi = \cos i/\sin\theta.$$

Therefore $$F_n = -\frac{3GM}{a_s^4} a_e^2 J_2 \sin u \sin i \cos i.$$

This force varies periodically round the orbit and is greatest for orbits inclined at 45° to the equator.

Substituting this expression for F_n in the Lagrangian equation for the node, the rate of change of the node in a circular orbit, for which $\eta = 1$, is found to be

$$\dot{\Omega} = -\frac{3GMa_e^2}{na_s^5} J_2 \cos i \sin^2 u.$$

$\dot{\Omega}$ also varies round the orbit but since

$$\sin^2 u = \tfrac{1}{2} - \tfrac{1}{2}\cos 2u,$$

it ranges between zero and a maximum equal to twice the mean value and with a speed twice that of the satellite in the orbit. $\dot{\Omega}$ is zero when u is zero or π, that is at the nodes, and attains its maxima when u is $\pi/2$ or $3\pi/2$, namely, at the points of maximum or minimum longitude.

The average rate of change is Ω is

$$-\frac{3}{2}\left(\frac{a_e}{a_s}\right)^2 nJ_2 \cos i.$$

The behaviour of the plane of the orbit under the force F_n is like that of a gyroscope.

The effects of all harmonic terms in the potential can be found from the equations for the variation of the elliptic elements and the results may be summarized as follows. The harmonics are classified into *zonal harmonics* that are independent of longitude, and *tesseral* and *sectorial harmonics* that depend on the longitude. The zonal harmonics may be further classified as *even*, that is those proportional to a Legendre coefficient with even powers of $\cos\theta$, and *odd* harmonics proportional to Legendre coefficients with odd powers of cos θ. The potential is dominated by the second order zonal harmonic which, as has already been seen, corresponds to the flattening and rotation of the Earth. Zonal harmonic terms of even order give rise to steady changes in the longitudes of the node and perigee as well as to small periodic terms in these and other elements which are negligible in nearly circular orbits. In particular, the node of a close satellite rotates at a rate of about

$$-\tfrac{3}{2}nJ_2\left(\frac{a_e}{a_s}\right)^2 \cos i,$$

or nearly 6° per day for a satellite with an inclination of 45°; the longitude of perigee of such a satellite changes at the rate of 6·5° per day. So far as zonal harmonics are concerned, the forces on a satellite do not depend on the position of the node because everything is symmetrical about the polar axis, but given an elliptical orbit, the forces do depend on the position of perigee, and therefore the rotation of perigee caused by the second order zonal harmonic term in the potential leads to variations in all the elements that vary periodically with the position of perigee. Thus in particular, the zonal harmonics of odd order give rise to variations of eccentricity and inclination proportional to the cosine or sine of ω. The forces that arise from tesseral or sectorial harmonics, being functions of the longitude, do depend on the position of the node of the orbit which, relative to the Earth, rotates nearly once a day; the effects of these harmonics therefore vary with a period nearly equal to a day.

The steady changes in the elements of an orbit are the easiest to determine and in particular, the rate of rotation of the node can be found simply from observations of the position of a satellite among the stars made with binoculars, so that it is not surprising that it was possible to make estimates of the coefficient J_2 as soon as the first Russian satellites had been launched; these early observations already showed that the value of J_2 previously estimated from geometric measurements and surface gravity was appreciably in error. Other elements are somewhat more difficult to determine by elementary methods of observation with the accuracy that is necessary to make reliable estimates of terms in the potential, but can be found with ample accuracy from the very precise means of tracking satellites and determining their orbits that are now used. The difficulties that exist in determining the harmonic terms in the potential do not come from the errors of the observations so much as from the fact that there are very many terms that have all much the same magnitude. The variation of a particular element, the steady change of the node for example, may be written as the sum of terms arising from all the harmonics that may contribute:

$$\delta\Omega_s = a_2J_2 + a_4J_4 + \ldots + a_{2n}J_{2n} + \ldots.$$

In this expression, $\delta\Omega_s$ is the observed change of the node after applying appropriate corrections and the coefficients a_n depend on the elements of the orbit—for example, a_2 is

$$-\frac{3}{2}\left(\frac{a_e}{a_s}\right)^2 \cos i$$

for a circular orbit. There is one such equation for each satellite. It might be thought that as a very large number of satellites has been placed in orbit, there should be no difficulty in obtaining sufficient equations to solve for a large number of harmonic coefficients. Unfortunately, that is not so. If two satellites are in nearly the same orbits, the equations for the change of the node will also be nearly the same. The two equations together will have a somewhat higher accuracy than either alone but will not enable any more harmonic coefficients to be found. The number of coefficients that can be estimated will not be more than the number of distinct orbits. It is not possible to divide orbits sharply into those that are distinct and those that are not, but if some of the orbits are nearly the same then although estimates may be made of a number of harmonic coefficients, many of those estimates may have very large uncertainties so that they will not differ significantly from zero. Now most close satellites, which are the ones that are most effective for the determination of the zonal harmonics of high order, have small eccentricities and as the coefficients do not depend strongly on e, any orbits that differ only in the eccentricities are for this purpose identical. The coefficients are proportional to $(a_e/a_s)^n$ and because a_e and a_s are very nearly the same for close satellites, most orbits that differ only in the semi-major axis are equivalent. Variations in the inclination of an orbit do, however, lead to great differences between the coefficients a_n and so the only really distinct orbits are those with different inclinations. Now the number of distinct inclinations is limited, because of safety and other restrictions at launching sites, to some seven values and thus the highest even zonal harmonic coefficient that has been estimated is J_{14}. But in making that estimate, all higher harmonics have had to be ignored and as there is no reason for thinking that they may be much less than J_{14}, it is clear that the estimates of J_{14} and harmonics of lower order may be appreciably in error. There is reason for thinking that the estimates of J_2, J_4 and J_6 are reliable because they have been found in different ways and in particular, it has been possible to determine them from orbits with semi-major axes great enough for the contributions of higher harmonics to be ignored. The estimates of J_8 and higher coefficients are much less secure. The terms which change with the period of the revolution of perigee also arise from a series of harmonics which contains a very large number of terms of which it is possible to estimate only a few.

The means used to find the tesseral and sectorial harmonics are more complex than those by which the zonal harmonics are estimated. These harmonics give rise to changes in the elements of an orbit that are proportional to the cosine or sine of $m(\lambda - \Omega)$ where, as before λ is the longitude of the satellite and Ω is that of the node. Now the measured longitude of a satellite depends on that of the place from which it is observed (elements such as the eccentricity, the inclination, the longitude of the node or the semi-major axis can be found without knowing the co-ordinates of the observatory) and therefore the harmonics cannot be determined apart from the coordinates of the observatory. In addition, the changes in the orbit due to the tesseral and sectorial harmonics are very small; these harmonics can thus only be found from the most accurate determinations of the orbit, in practice from observations with the Baker-Nunn cameras of the Smithsonian Astrophysical Observatory with which the direction of a satellite can be measured to about 1 second of arc. In working out the results it must be supposed that the positions of the cameras are not precisely known and that both the positions as well as the harmonic terms in the potential have to be derived from the observations. For this reason and because, as with the zonal harmonics, any particular change in an element of the orbit can in principle arise from an infinite series of harmonic terms, for example, all those proportional to $P_n^m(\cos \theta)$ where n is variable but m takes a specific value, the estimates of the tesseral and sectorial harmonics cannot be considered to be very reliable. A second way of finding tesseral and sectorial harmonics has been employed recently. If the period of a satellite in its orbit is a simple rational multiple of the period of the rotation of the Earth about its polar axis, the forces on the satellite will repeat after some complete number of revolutions of the satellite and the elements may show steady changes or, in practice, since the relation of the period of the satellite to the day is not likely to be exact, changes with a very long period. Such effects will only be seen with satellites that are far from the Earth. Thus for the satellite to have a period of one day, it must be at a distance of about $6\frac{1}{2}$ Earth's radii (a geostationary satellite) and observations of slow changes in the orbits of geostationary satellites enable the second order tesseral harmonic corresponding to an elliptical equator to be estimated. If the satellite is closer to the Earth, the relation between its period and the length of the day will be more complex, but effects arising from terms proportional to harmonics with an equal to 12, 13 and 14 have been detected and estimates have been made of the coefficients of those harmonics. It must be emphasized once again that, as with all other methods of estimating harmonics from the behaviour of satellites, it is only possible to find the first terms of a series that is in principle infinite.

Current estimates of coefficients of zonal harmonics in the external potential are listed in Table 1. Values of uncertainties have been given only for the zonal harmonics of lowest order because it is for these alone that there is any confidence that they are relatively unaffected by neglect of higher undeterminable harmonics. All other terms should be regarded as little more than guides to the orders of magnitude of the terms themselves and of the next few terms in the series. Great confidence can however be placed in the actual values of J_2, J_3, J_4, J_5 and J_6 for these have varied little

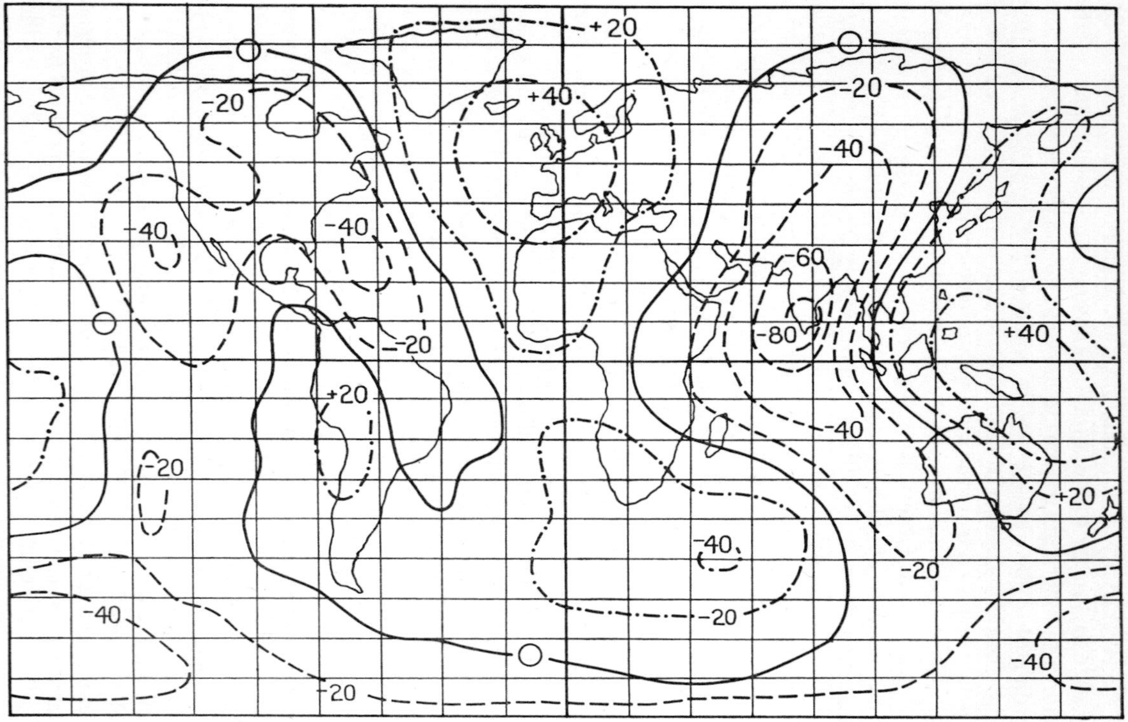

Fig 12 Map of the geoid. Figures show the height of the geoid in metres above a spheroid of flattening 1/298·25

TABLE 1

Values of coefficients of zonal harmonics in the external gravitational potential

n	$10^6 J_n$
2	1,082·65
3	−2·54
4	−1·61
5	−0·25
6	0·71
7	−0·4
8	0·0
9	0·0
10	−0·2
11	−0·0
12	−0·1
13	0·0
14	0·2
15	−0·2
17	0·0
19	0·0
21	0·2

with the number of additional harmonics that have been included, and in particular the flattening of the Earth can be considered to be known with an accuracy nearly one hundred times greater than it was before artificial satellites were launched. The results of the determination of the harmonics may be shown alternatively as a map of the geoid, for the corresponding terms in the radius vector of the geoid can be found from the relation given in the previous section and the departure of the geoid from a spheroid having the flattening corresponding to a particular value of J_2 may be shown as a contour map such as that reproduced in Figure 12. While the large scale features of such maps, those up to harmonics of order 5 or 6 are probably fairly reliable, no great weight should be attached to details of higher order for the reasons just explained.

Newton realized that gravity should vary over the surface of the Earth and measurements of that variation have been made from his time onwards. Up to about 1940, measurements were made almost entirely with pendulums, were very scattered and apart from some small special areas, were confined to the land. The development of spring balance gravity meters for geophysical prospecting made it possible to make surveys on land very quickly, but measurements at sea remained few so long as they could only be made with pendulums in submarines. The difficulty with observations at sea is that the observed value of gravity includes the instantaneous acceleration of the ship due to its motions on the waves. Pendulums respond non-linearly to the acceleration and the average result is thus not free of the effects of the periodic wave motions; for that reason, pendulums must be used in submarines which can operate at depths where the surface wave motion is largely ineffective. Technical improvements in gravity meters and in the methods of recording and reducing their readings made it possible to use them in surface ships, and gravity surveys of the oceans were then greatly accelerated, mainly because it is much easier to obtain the use of a surface ship than of a submarine, but also because it is easier to navigate from a surface ship. Navigation is the major source of error in gravity measurements at sea. The observed values of gravity have to be compared with the values calculated from some conventional formula and to obtain sufficient accuracy in that comparison, the latitude should be known to about half a nautical mile. This is itself sufficiently difficult in the deep oceans but a greater difficulty is presented by the need to know the ship's speed with high precision. An east-west component of the speed of the ship is equivalent to a change in the spin velocity of the Earth and thus to a change in the observed value of gravity according to the formulae given in the preceding section. To correct for this effect to 1 part in a million of gravity, the east-west speed of a ship on the equator should be known to about one-tenth of a knot, an accuracy that is at present impossible. The speed of the ship through the water can of course be found with such an accuracy, but far from land the ocean currents are uncertain by much greater amounts. Conventional navigational methods are quite inadequate but navigation with the help of special artificial satellites may give improved precision.

The seas occupy about three quarters of the surface of the Earth and not only are gravity measurements over them far less accurate than those on land but also by far the larger part remains without any observations. Thus in attempting to express the surface value of gravity as a series of surface harmonics, it is found that the variation of gravity is but poorly sampled, for the most accurate observations (those with an uncertainty less than one part in a million) are confined to land areas and the area of the sea that is surveyed is still relatively small. It is accordingly not surprising that the harmonics that can be determined in the series for the surface value of gravity are many less than can be determined in the potential. Those that can be determined do not agree well with the harmonics in the external potential and while there may be some theoretical uncertainty in comparing the one series with the other, arising from the difficulty that the geoid, for example is not a surface wholly outside the matter of the topography, nonetheless, it is probable that the inconsistencies reflect the poor sampling of surface gravity.

For these reasons, the accuracy of the estimate of J_2 from surface gravity is far less than that from the external potential. On the other hand, the local variation of gravity in a restricted area cannot be obtained from satellite data because the motion of a satellite is not sensitive to the high harmonics, say of order 50 or more, that are needed to represent the local variations of gravity in an area about 1000 km across, whereas local gravity surveys can give those variations with good accuracy.

If it is desired to determine the large scale parameters of the Earth, such as the polar flattening or the ellipticity of the equator, the results from the observations of artificial satellites will be used, as will they be if the purpose is to relate surveys in areas separated by wide oceans; in these cases, use will be made of the series of harmonics derived from the satellite data. If, however, it is required to know the local form of the geoid so as to relate surveys that are separated by perhaps 100 or 200 km, then surface measurements of gravity will be used and the variations of the radius vector of the geoid or of the direction of the normal to the geoid will be found from gravity data measured over and between the areas of the surveys, and the calculations will be made with local integrals of these gravity values and not by means of an harmonic series supposed to represent the world wide variation of gravity.

2.3 Modern Survey Methods

The last two decades or so have seen the development of methods for the measurement of distance depending on the time of travel of electromagnetic waves and the consequent replacement of geometrical methods of survey based on the measurements of angles by those based on the measurements of length, with considerable improvements in the convenience and accuracy of surveying. It has already been pointed out that the accuracy of a network of triangles is much greater if all the sides are measured, than if all the angles and just one or two of the sides are measured. Electro-magnetic methods now cover a range from a few tens of metres to hundreds of kilometres and are applicable to a wide variety of problems. Broadly speaking, radio waves are used for the longer distances and light for the shorter.

There are features common to all electromagnetic methods. In all accurate schemes, the phase of an oscillation is measured, either of a carrier signal itself or of a modulation imposed upon it. Depending on the techniques employed, phase can be measured to between one hundredth and one ten thousandth of a wavelength and the wavelength of the carrier or modulation signal as the case may be must be chosen so that the least distinguishable phase difference corresponds to the accuracy required of the measurement. This will lead to a wavelength that is a small fraction of the distance to be measured—for example, a microwave system with a carrier wavelength of 10 cm and phase measurable to a few millimetres might be used to measure lengths of 20 km, or two hundred thousand wavelengths. The measurements will be ambiguous unless the distances are already known to within one wavelength and means must therefore be provided for resolving the ambiguity. The usual way in to use a number of different frequencies or wavelengths; the excess fractions of a wavelength in the distance, which are the quantities actually measured, will only be consistent if the correct whole numbers of wavelengths are assumed. A second problem common to all electromagnetic methods is that the effective wavelength depends not only on the frequency, which can be measured very accurately, but also on the refractive index of the air for the carrier wave, whether it be radio or optical. The relation of refractive index to the density and composition of the air has been determined very accurately for both radio and optical wavelengths but much of the path along which the signal travels is inaccessible so that it is difficult to obtain the mean density and composition of the air for the effective wavelength to be calculated to better than one part in one hundred thousand. The usual practice is to measure the pressure, temperature and humidity at the two ends of the line being measured and to assume that the mean of the two sets adequately represents the mean conditions along the path. That may be so in favourable circumstances, but there are many conditions in which it will not be so, for example if surfaces of constant temperature follow the ground and the two ends of the line are at a considerable height above the intervening ground. The refractivity changes by about one part in a million for one degree Celsius change of temperature and for about 3mb change of pressure so that it should usually not be difficult to obtain values that are correct to one part in a hundred thousand, but only rarely will it be possible to obtain one part in a million without elaborate sampling of the air path. As has already been said, one part in a hundred thousand is less than has been obtained in the very best measurements of base lines with invar tapes, but is far better than the accuracy of scale in a network of triangles with no more than one or two base line measurements. Thus electromagnetic measurements with quite simple estimates of refractive index are a great improvement on triangulation with angular measurements. Electromagnetic measurements are also much more convenient, for fewer observations are required than with theodolites in which complex sequences of observations are required to enable various instrumental errors to be eliminated, and further, electromagnetic methods are generally less dependent on weather conditions than theodolite observations which require the best atmospheric optical conditions.

Stations up to 30 km apart are usually intervisible in most parts of the world and distances between them can be measured with microwave radio waves or with modulated optical waves. The *Tellurometer* is the most widely used and most effective microwave system. A difficulty with radio systems is that if they are to be transportable, the aerial system must be small, and so only a few wavelengths across, so that the radio beam spreads out widely from it and if a simple reflector, again of small dimensions, is placed at the far end of the line to be measured the intensity of the reflected beam returned to the transmitter will be very small. Radar systems can accept such a reduction because the peak power in a pulse of energy can be very great, but time measurements on pulses are not accurate enough for geodetic survey, so that continuous signals, of which the power is necessarily much less, have to be used. The difficulty is overcome in the Tellurometer by placing at the distant end a second transmitter controlled by the signal received from the first. With a simple system of a receiver followed by a transmitter, there would be a somewhat indeterminate difference of phase between the received and re-transmitted signals and in the Tellurometer there are arrangements to eliminate that uncertainty. A source of error with the Tellurometer is that the wavelength is not very small compared with the distance that the beam may pass above the ground and therefore effects of diffraction and of stray reflections may be important; they can be estimated from the change in apparent length when the frequency is altered slightly.

The *Geodimeter* is an example of an instrument using a light beam. An intense beam of light is passed through an amplitude modulator and then transmitted from a collimator to a reflector at the distant station; on reflection to the transmitter it is received at a telescope and the phase of the modulation is compared with that of the transmitted beam. The modulation is achieved by passing the light through a polarizer to make it plane polarized and then through a Kerr cell containing nitrobenzene which rotates the plane of polarization when a high voltage is applied to it. On passing through a second polarizer, the amplitude of the light varies at twice the frequency of the voltage applied to the Kerr cell. The reflected signal falls on the cathode of a photomultiplier to which is applied a voltage in phase with the modulating voltage applied to the Kerr cell: if the phase of the returned signal is the same as that of the modulating voltage, the current from the cathode will be amplified, but if it is in opposite phase, it will not be amplified. The signal from the photomultiplier is therefore a measure of the difference of phase between the transmitted and received signals. Measurements with the Geodimeter are not subject to diffraction or stray reflection errors, but they do depend on the colour of the light since the refractive index of air varies slightly with wavelength; the effective mean wavelength, which depends on the response of the photocathode, must be determined with some care.

The instrumental errors of the Tellurometer and the Geodimeter are each a few millimetres, adequate for measurements of lengths of 20 or 30 km for which the errors due to uncertainties of refractive index may be a few centimetres, but too great if measurements are to be made over 2 or 3 km. Such distances do not often arise in first order geodetic measurements but may occur in special applications, for example in studies of possible changes in lengths set out on the ground. Such studies are on the border line of geodesy and are required both for large scale engineering work and for scientific purposes. The *Mekometer* is an instrument with much smaller instrumental errors than the Tellurometer or the Geodimeter, especially suitable for these shorter distances. It also uses a modulated beam of light but, unlike the Geodimeter, employs polarization modulation, not amplitude modulation. Two advantages flow from this choice. In the first place, changes in optical properties of the air along the path change the amplitude of a reflected signal but do not alter the polarization and so the noise on the indication of the phase difference between the transmitted and the received signals is much less with polarization modulation than with amplitude modulation. In the second place, the method of detection is more suitable. In the Mekometer, a plane polarized beam of light is passed through an optically active crystal of potassium dihydrogen phosphate which rotates the plane of polarization when an electric field is applied to it. The reflected light is redirected through the same crystal and if the phase is such that after rotation in the crystal, the plane of polarization is parallel to that of the polarizer that produced the original plane polarized light, the reflected light returned through the polarizer will have maximum intensity. The intensity of the light as indicated by the output of a photomultiplier is thus, as in the Geodimeter, a measure of the phase difference between the transmitted and reflected light. There is however a considerable difference in the behaviour of the photomultiplier as a detector. In the Mekometer, the amplitude of the light is not varying—for given distance of the reflector, the light falling on the photocathode has a constant intensity, determined by the phase of the returned polarization. The high frequency behaviour of the photomultiplier therefore does not affect the behaviour of the instrument, nor does it limit the frequency of modulation. In the Geodimeter, on the other hand, the amplitude of the light falling on the photocathode varies at the modulation frequency and the response of the detector depends on the relation of the transit time of electrons in the photomultiplier to the period of the modulation. The highest modulation frequency that can be used in the Geodimeter is limited by the transit time effect in the photomultiplier as well as by the highest frequency that can be applied to the Kerr cell without the electrical losses becoming too great. Thus whereas the Geodimeter is limited to a frequency of 30 MHz and therefore to a modulation wavelength of 10 m, the properties of potassium dihydrogen phosphate would allow the Mekometer to work at a frequency in excess of 30 000 MHz or a wavelength less than 1 cm. Such short wavelengths are not often necessary, for with a modulation frequency of 500 MHz or a wavelength of 60 cm, the instrumental errors are not more than 0·2 mm.

Apart from considerations of accessibility, there is considerable advantage in surveying a region by means of measurements over as great distances as possible, because the possibilities of accumulations of errors are less the fewer the measurements needed to cover a given area. Optical observations between points on the surface, whether angular measurements or measurements of distance, are limited to distances of some 30 to 50 km.

The strongest survey networks are those consisting of a set of triangles and while it is no longer essential to adopt such networks because most of the measurements must be of angles, it is still true that the best results will be obtained with such nets, if only because it is then possible to dispense with angular measurements altogether. Economic considerations, however, may restrict trilateration surveys, as these may be called, to small regions of greatest economic value, for over large areas they will be very expensive. Large and less developed areas, for example in Africa or Australia, may therefore be surveyed by lines of traverse, in which survey points are related one to another by the distances between them and the bearing of one from another. The Tellurometer and the Geodimeter have made such precise traverse a relatively speedy and accurate procedure, as compared with earlier practice, for example in the USA, where precise traverse was conducted by measurements with tapes along railroad lines. The distances measured along lines of precise traverse by electro-magnetic methods are sufficiently accurate but the weakness of traverse is in the directional measurements in which errors may accumulate over long lines of traverse. Angular uncertainties can be controlled by astronomical measurements which must be made at sufficient places in the lines to prevent undue accumulation of errors.

As an alternative to, or a control on, precise traverse, it is possible to conduct trilateration on a very large scale if measurements of distance are made, not between points on the ground, but between such points and an aircraft at a great enough height to be visible from the ground sites. If the aircraft carries a radio transmitter, the distances between it and each of a number of ground stations may be found from simultaneous measurements of the phases of the signals received at the stations. Consider an aircraft flying at a height h at right angles to the line joining two ground stations, A and B (Figure 13). Let x and y be the distances from the aircraft to stations

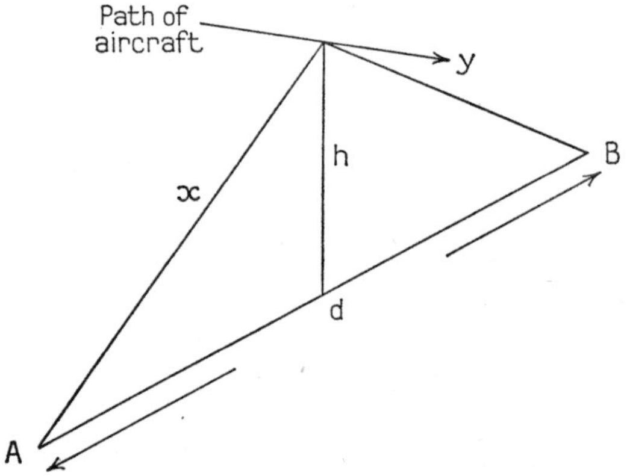

FIG 13 Line crossing measurements of distance

A and B respectively. The minimum values of x and y are recorded; they occur when the aircraft is above the line AB. Then

$$d = (x^2 + h^2)^{\frac{1}{2}} + (y^2 + h^2)^{\frac{1}{2}}.$$

From this formula it follows that

$$\frac{\partial d}{\partial h} = \frac{h}{(x^2 + h^2)^{\frac{1}{2}}} + \frac{h}{(y^2 + h^2)^{\frac{1}{2}}}.$$

It will be seen that, if for example, h is 1 km and x and y are each 200 km, the accuracy with which h needs to be determined is 1 per cent of the accuracy desired in the distance AB. h can be found to such accuracy with some form of electrical altimeter, for instance a capacity meter.

In practice, the height is somewhat indefinite, but the difference between the heights of two crossings of the line *AB* should be well defined and can be used instead.

The usual arrangement is for the ground stations to have radio transmitters and for the aircraft to carry an instrument (transponder) that receives the signal transmitted from the ground and retransmits a signal having a constant phase difference from that transmitted from the ground. This is like the Tellurometer but the frequencies are usually lower and the uncertainties of the phase shifts are greater, a situation that is quite acceptable over the very much longer lines that are observed. Various systems of differing versatility and precision are available, for instance, HIRAN, SHORAN and AERODIST. Over distances of about 300 km uncertainties may be found 1 to 3 m. Much shorter distances have been measured by a similar procedure from a helicopter.

The traditional geometrical methods of geodesy have been confined to measurements over land areas, and devising means to connect areas separated by seas has always exercised the ingenuity of geodesists. Early in the history of geodesy, England and France were connected by careful surveys

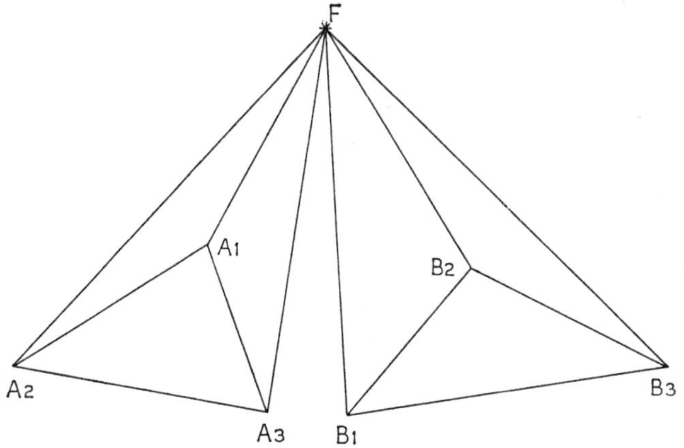

FIG 14 Geometry of flare triangulation

across the English Channel and it was in fact for this purpose, that the first geodetic base in England, that on Hounslow Heath, was established. Geometrical methods are limited by the need for sites to be intervisible and so no great progress could be made until balloons, rockets or aircraft were available to carry instruments to heights from which they were simultaneously visible from widely separated places on the ground. The measurement of distances from aircraft enables points some 300 km apart to be connected, but greater distances can be covered by observing the directions of lights against a background of stars, a procedure known as flare triangulation. The geometry of this technique is shown in Figure 14.

Let A_1, A_2, A_3, be three stations of which the relative coordinates are known from local geodetic survey. Let F be a flare and let B_1, B_2, B_3, be another set of three stations, also related by local geodetic survey. Then it is evident that if the directions A_1F, A_2F and A_3F are known, the position of F in the A coordinate system can be calculated. So, likewise, can the position in the B co-ordinate system and in this way the relation of the B to the A system may be found. The observations from the A and the B systems must be synchronized.

Angular accuracies of about 1 inch can be obtained, corresponding to errors of 2–3 m in horizontal distances. Flare triangulation has been used to connect Norway and Scotland.

Measurements between the ground and objects in the sky give valuable results for horizontal distances but are inadequate for accurate comparisons of heights for much the same reasons as conventional triangulations is unsuitable for measurements of heights—the angular accuracy of about 1 second of arc corresponds to 10 cm at 20 km as compared with an accuracy of a few millimetres that can be obtained with good spirit levelling, and the systematic errors of refraction with rays at 10 or 20° to the horizon are serious. The problem of comparing potentials across the seas is very far from satisfactorily solved. The level of the sea cannot itself be used because, especially in narrow seas, it is distorted by currents, although some attempts have been made to measure currents by the electrical potential they generate in the magnetic field of the Earth and to use the data so obtained to correct the observed mean sea level. Good results have been obtained across the English Channel, but the conditions are particularly favourable for the measurements. Another method, used between the islands of Denmark, has been to lay a tube with water in it under the sea and to observe the levels of the water at the two ends of the tube. When used in engineering, such water levels have proved rather difficult to operate because of the effects of differences of temperature along the tube and of differences of atmospheric pressure at the ends, and the same difficulties occur in geodetic work. Again, the laying of such tubes is practically restricted to narrow seas or channels.

Photogrammetry is almost universally used for mapping nowadays and is usually based on points that have been surveyed by the geodetic techniques that have just been described. It is possible with great care to use photogrammetric observations for geodetic work but the scale has to be obtained from bases laid out on the ground.

Geodetic survey of land areas is now, thanks, largely to electromagnetic methods of measuring distance, convenient and accurate but it is still a matter of some difficulty to connect such areas even across narrow seas, to say nothing of the wide oceans, by geometrical methods, while present knowledge of the gravity field of the Earth is too partial for extensive use to be made of the inferred form of the geoid. Despite the great advances made in techniques of geometrical measurements and in the applications of satellite observations and gravity measurements, the programme outlined earlier of referring local surveys to the equipotential surfaces as local reference surfaces and then comparing them through the known properties of the equipotential surfaces, is far from complete and so great interest attaches to another application of artificial satellites, the direct measurement of distances between points on the Earth that enables the idea of referring local surveys to a world-wide polyhedral frame to be revived.

Geometrical observations with artificial satellites

The geometrical measurements that can be made with artificial satellites may be used to obtain relations between groups of points on the surface of the Earth and so to build up a polyhedron embracing the Earth or they may be used to give the geocentric coordinates of points on the surface. In the first application, a satellite is observed simultaneously from a number of points on the Earth and it is considered simply as a light in the sky as in flare triangulation, although the types of observation that may be made are more varied than with flare triangulation and the height of the satellite can be very much greater than an aircraft, a balloon or a rocket. No use is made of the knowledge that the satellite is in an orbit to relate successive positions of the satellite one to the other. In the other application, the positions of the points from which the satellite is observed are related to the orbit of the satellite. The first procedure is independent of the external gravitational field of the Earth but in the latter, the orbit and the positions of the stations are interdependent.

The geometry of the simultaneous observations is the same as that for flare triangulation shown in Figure 14. The main practical differences from flare triangulation are that the satellite is moving relatively rapidly across the sky instead of drifting nearly stationary in the atmosphere, that as a light source it is much weaker because far more distant, and that since the ground stations are more distant, the synchronization of the observations calls for more care. Passive and active satellites have been proposed and used. The Echo balloons were passive satellites, inflated balloons in orbit at heights of about 1,000 km, which shone by the reflected light of the Sun. Observations are restricted by two conditions, that the satellite shall be in a position where it is visible from the two areas to be connected, and that it shall be in sunlight above the shadow of the Earth while at the same time each of the observing regions shall be in darkness. The Echo satellites were fairly bright and could be photographed with cameras of rather small aperture, but they were large, some 60 m in diameter, and so the uncertainty of coordinate measurements cannot be much less. The synchronisation of observations of passive satellites that shine continuously in the reflected light of the Sun depends on synchronized time signals at the ground sites and therefore on reliable time service and standard frequency broadcasts. Active satellites are those that carry a flashing light or other signal and the observations are necessarily synchronous, subject only to the condition that the flashes are correctly identified on the records. In some ways, however, they need more elaborate ground services. The bright Echo satellite could readily be found from quite coarse predictions of its orbit and its trace, interrupted by a shutter in the camera synchronized with the time signals, could easily be located on the photographic plate. The flashing light satellite visible only for a very short time, cannot be found without very accurate predictions of the orbit and accurate setting of the camera. On the other hand, the active satellite has the advantage that it can be seen from any place in darkness and not only when it is illuminated by the Sun. So far almost all geodetic measurements have been made by measuring the direction of a satellite against the stars from a photograph, but measurements may also be made of the distances of the satellite from ground sites. The trials that have been carried out on the measurement of distance have made use of solid state pulsed lasers, the time of travel of a pulse of light from the ground to the satellite and back being recorded.

In the alternative way of using satellites, knowledge of the orbit is exploited. The direction of a satellite from some site is observed and then later, its direction from another site, and the two sets of observations are connected through the knowledge of the intervening orbit. If the orbit is known, the position of a ground station with respect to it may be found and so all ground stations may be related to the orbit. If the orbit is known with respect to the centre and polar axis of the Earth, the stations can be assigned geocentric coordinates. The advantage of this scheme is that stations do not have to be chosen so that they are simultaneously visible from a given satellite position yet are in darkness while the satellite is illuminated. Thus greater distance between ground sites and a greater choice of sites is possible. The disadvantage is that the accuracy with which the orbit may be determined is restricted and that the determination of the orbit is linked with that of the positions of the ground sites. As was explained earlier, the determination of changes in the orbit brought about by tesseral or sectorial harmonics in the potential depends on knowledge of the positions of the sites and so the harmonics and the station positions have to be found simultaneously from a least squares reduction of the data. At present, there are few sites (those operated by the Smithsonian Astrophysical Observatory) and as explained above, few distinct satellite orbits, so that the normal equations for the parameters are very poorly conditioned and in consequence the estimates of positions and harmonics have relatively large uncertainties and are strongly correlated. A further source of error lies in the drag of the air, for while that can be allowed for on broad lines, it is rather erratic and cannot be predicted with high accuracy from one revolution to the next or even from one part of the orbit to another. As a result, not only are the accuracies of positions along the orbit limited, but also, it is difficult to obtain predictions of the accuracy desirable for laying the cameras used in these observations. Despite these difficulties the work of the group at the Smithsonian Astrophysical Observatory has produced the only significant geometric results so far since very little has yet been accomplished by simultaneous observations.

The Moon as a satellite of the Earth is an object that can be observed to provide geometrical positions on the Earth. The direction of the Moon from a place at the surface of the Earth can be found from the directions of stars as they pass behind or are uncovered by the Moon and from such observations (or occulations), the geocentric coordinates of the observatory may be found. The accuracy of occultation observations is much less than that of observations of artificial satellites mainly because the Moon moves much more slowly across the sky. The possibility of placing objects on the Moon together with that of measuring distances to the Moon by means of pulsed lasers led to the suggestion that a cube corner reflector could be placed on the Moon and that its distance from a site on the Earth could be measured by timing laser pulses reflected from it. A remarkable range of data might be obtained from such measurements, in particular geocentric coordinates with an accuracy that would enable variations of position of a few centimetres to be detected. A cube corner reflector was placed on the Moon by the Apollo 11 crew, and laser reflections have been observed.

Artificial satellites, the facilities afforded by space probes and the intense radiation from lasers are together beginning to make it possible to overcome the restrictions imposed by the curvature of the Earth and the refractive index of the atmosphere that have hitherto prevented the determination of the form of the surface of the Earth by purely geometrical methods and while only preliminary and inadequate results are yet available, it is clearly possible to foresee that the geometrical determination of the form of the solid surface of the Earth will become independent of the determination of the external gravitational potential. It is in fact possible to foresee the time when the external potential, the surface value of gravity and the geometrical form will all be found independently and the one will be checked by the others instead of as now, the determination of one depending upon the others.

2.4 Geophysical Data and the Stability of the Surface of the Earth

The shape of the equipotential surfaces, the external potential and the surface value of gravity may all be expressed in terms of the coefficients J_{nm} and in particular, the flattening and the part of the variation of gravity dependent on $\cos^2\theta$ may both be written in terms of the coefficient J_2 in the potential. This coefficient is related to the moments of inertia of the Earth:

$$J_2 = C - \frac{\frac{1}{2}(A + B)}{Ma_e^2}$$

where A and B are the principal moments of inertia about axes in the equatorial plane and C is the moment about the polar axis. Now the rate of precession of the axis of the Earth about a direction fixed in space due to the attraction of the Sun and the Moon is proportional to the ratio

$$H = \frac{C - \frac{1}{2}(A + B)}{C}$$

and thus from the two parameters J and H, both of which are well determined from observation, it is possible to calculate the moments of inertia. The values are

$$C/Ma^2 = 0{\cdot}33085$$
$$A/Ma^2 = 0{\cdot}32977$$

assuming, as it very close to the truth, that A and B are equal.

These results show at once that the Earth is strongly condensed toward the centre. If the Earth were a sphere of uniform density, the ratios C/Ma^2 and A/Ma^2 would each be equal to 0·4. As the ratios are less than 0·4 the central density must be greater than the superficial density. If we take a simple model of a sphere having a density ρ_1 in the inner region below a depth equal to half the surface radius and ρ_2 in the outer region, we find that

$$C/Ma^2 = 0{\cdot}4 \times \frac{1 + \dfrac{\rho_1 - \rho_2}{32\rho_2}}{1 + \dfrac{\rho_1 - \rho_2}{8\rho_2}},$$

and if $C/Ma^2 = 0{\cdot}331$, $\rho_1 = 2{\cdot}4\rho_2$; a relatively small difference of C/Ma^2 from 0·4 implies a strong central condensation.

The values of the moments of inertia are used in the calculation of densities within the Earth from seismic data. If α is the velocity of longitudinal waves and β that of shear waves in the Earth, the observed times of the travel of earthquake waves from point to point across the surface of the Earth give α and β as functions of geocentric radius. Now

$$\alpha^2 - \tfrac{4}{3}\beta^2 = k/\rho$$

where k is the bulk modulus of the material of the Earth and ρ is the density. These results may be combined with the definition of k:

$$\frac{k}{\rho} = \frac{dp}{d\rho}$$

(p is the hydrostatic pressure): with the differential equation for the pressure:

$$\frac{dp}{dr} = -g\rho$$

and with the expression for gravity at a radius r:

$$g = Gm/r^2,$$

where m is the mass within the sphere of radius r, to give an equation:

$$\frac{d\rho}{dr} = -\frac{Gm\rho}{r^2(\alpha^2 - \tfrac{4}{3}\beta^2)}$$

from which ρ may be found as a function of r provided the constants of integration can be determined. Were the Earth to show no discontinuities of seismic velocity, one constant only would need to be determined—the total mass—but, since there are major discontinuities, especially that between the core and the mantle at about half the surface radius, it is necessary to use other data if an unique solution is to be found. The polar moment is the second important constant of integration.

At first sight, the mass of the Earth could be found from the observations of artificial satellites, since according to Kepler's laws, the period T_s and the semi-major axis a_s, of an orbit are related by the equation

$$GM = 4\pi^2 a_s^3/T_s^2.$$

It is not, however, possible to find M, or rather, the product GM, from the orbits of close satellites with high accuracy, because the resistance of the atmosphere, to which close satellites are subject, changes the period from that given by Kepler's equation. Distant satellites are not subject to air drag and in particular it should be possible to find the mass of the Earth from the orbit of the Moon. The period of the Moon in her orbit is known very exactly and in recent years the distance between the Earth and the Moon has been measured by radar, although there is some uncertainty in deriving the distance to her centre of mass from the measured distance to the nearest point on the surface of the Moon. There are two other complications. The attraction of the Sun has a considerable influence on the motion of the Moon and the observed period of the Moon in her orbit; and the mass of the Moon is not so small in relation to that of the Earth that it may be ignored in calculating the period. The theory of the effect of the Sun on the orbit of the Moon is well established and the effect can be calculated with great accuracy, while the mass of the Moon has been determined very precisely from its effect on the apparent motion of a space probe. The principal uncertainty in determining the mass of the Earth from the orbit of the Moon therefore lies in the geometrical uncertainty of the position of the centre of mass of the Moon in relation to the radar observations.

The product GM may also be found from the value of gravity at the surface of the Earth, supposing a_e to be known. The uncertainty is appreciable however, mainly because of the poor sampling of the surface of the Earth with gravity measurements. The mean value of gravity over the surface is found by measuring the value at a few points in absolute terms, that is directly in relation to the units of length and time, and then relating these measurements to the mean value by measurements of differences of gravity covering the surface. Just as the poor sampling of the surface by these differential measurements distorts the value of J_2 derived from surface gravity measurements, so does it distort the mean value of gravity.

The best determinations of GM come from observations of the accelerations and distances of artificial satellites or space probes at great distances from the Earth and the best results are obtained when the observations are made by means of the Doppler shifts of the frequency of radio transmissions from the satellite or space probe, observations which give directly and with very high accuracy, the velocity of the object relative to the Earth.

This brief review of current work on the major mechanical parameters of the Earth shews that the mass and moments of inertia can be derived from observations made solely on the surface, but that the most satisfactory results depend on observations of the Moon, artificial satellites and space probes. The same is to some extent true of the equatorial radius of the Earth. The speed n_m of the Moon's motion in her orbit is related to the semi-major axis of her orbit a_m, to the mean value of gravity on the surface of the Earth g_e, and to the equatorial radius of the Earth a_e, by the following equation, which has been simplified by omitting small additional terms depending on the polar flattening of the Earth, the mass of the Moon and the attraction of the Sun on the Moon:

$$n_m a_m^3 = g_e a_e^2.$$

Evidently, a_e may be found if the other quantities are known. Since in practice it is possible to measure all the quantities involved, it is best to use this equation as a condition to be satisfied by estimates of the parameters, and in 1964, the International Astronomical Union adopted a set of constants that were consistent in this sense and were based on the most reliable observations then available.

While the mass and moments of inertia of the Earth enter the determination of the variation of density with radius within the Earth and thus contribute to knowledge of the major structural divisions of the Earth and to the equations of state that apply within them, the higher order variations of surface gravity and the external potential give some information about the variability of density with angular position, and about the depths at which such variations may occur. Since the potential of a uniform spherical shell is the same as that of a point mass at the centre, it is only because there are angular variations of potential that we can learn anything from gravity about the distribution of density with depth. This is true equally of the moments of inertia, for it will be observed that the quantity found from the gravitational parameter J_2 is the *difference* of the principal moments of inertia. The inequalities of density that give rise to the lowest harmonics in the potential could be put anywhere in the Earth but the existence of harmonics of order greater than the twelfth shows that there must be inequalities in the mantle of the Earth since if supposed to lie at greater depths, they could not give rise to detectable harmonics of this order unless the variations of density were quite improbably large.

On a much smaller scale, it is found that the surface values of gravity are correlated with topography in such a way as to imply that excess matter at the surface is associated with a deficiency of matter at a depth of some 30 km a principle that is found to hold quite generally and that is known as the principle of *isostasy*. Isostasy seems to be applicable only to the crust. Seismic studies on land and at sea have made it possible to establish the variation of density with depth under the continents and under the oceans so that the attraction of gravity due to the continents and oceans may be calculated. The values so obtained are very close to those observed and agree closely with the principle of isostasy but the residual variation of gravity does not seem to be correlated with the continents and oceans on a world wide scale, implying, apparently, that the structure of the continents and oceans, in the large, is not reflected in the mantle beneath. This conclusion, like that depending on correlation with heat flow, is perhaps not too secure for it depends on the detailed values, rather than the orders of magnitude, of the harmonics in the external potential of orders from about 4 to 12 and, as was emphasized above, the actual values of harmonics beyond the 6th are not well known in detail. Furthermore, the harmonics which are best determined are not those of most importance in expressing the surface distribution of continents and oceans. Nonetheless, the conclusions to be drawn from the data at present available are that there are appreciable variations of density within the upper part of the mantle on a world-wide scale that do not appear to be directly related to the distribution of continents and oceans, but which do appear to be related to the heat flow through the surface of the Earth.

So far the determination of the geometrical and mechanical properties of the Earth has been discussed as if the Earth showed no change with time. In detail that is not true. Geometrical relations are changed violently by earthquakes and historical and prehistorical changes of level provide evidence of slower changes .The methods of geodesy are some cases now sensitive enough to detect such changes unambiguously.

Earthquake movements can of course be readily seen from the gross, abrupt changes that they produce at the surface, but there is considerable interest in trying to detect the development of strain over a wide area prior to the occurrence of an earthquake, and the changes of strain subsequent to an earthquake. Conventional methods of triangulation have revealed these slow changes in California where there is great and continuous seismic activity, and in the neighbourhood of the large faults it is sometimes possible to detect changes in the course of less than a year through the displacements of drill pipes in oilfields. There is a great need to be able to detect movements in less seismic areas and to measure them in more detail in the more active regions, and electromagnetic methods of measuring short distances, in particular, the Mekometer, seem capable of the necessary accuracy of a few tenths of a millimetre and has already been applied in such studies. More delicate measurements require optical interference methods and while methods using gas lasers are being actively developed, they are outside the scope of geodesy strictly speaking.

Changes of height were probably the first widespread changes to be detected geodetically. Comparisons of geodetic levelling over periods of a decade or so have shown that the east coast of Britain is sinking while the west coast is rising. Similar movements have been detected elsewhere, but the results still need to be treated with caution since earlier schemes of geodetic levelling were subject to systematic errors comparable with the changes thought to have occurred. It is certain that changes of level have occurred and some of the better established instances are in regions, such as the British Isles and the Baltic Shield, that are otherwise stable. Some changes of level, as in Canada and the northern parts of the USA and in Scandinavia, are due to the removal of the load of the recent ice caps, but there are others that do not seem to be explicable in that way.

Absolute methods for the measurement of gravity are scarcely reliable enough at present to detect changes in values of gravity, but very sensitive measurements of differences do reveal variations of gravity from site to site that appear to be correlated over quite wide areas. These studies are still in their infancy and the significance of the results in terms of tectonics or otherwise, are not yet apparent. In volcanic regions, of course, very local changes of gravity may arise from movements of magma.

Changes of level or of gravity from place to place imply changes of the direction of the vertical relative to the fixed stars and relative to the surface of the Earth. The former appears as a change in the astronomical coordinates of a place and the dominant part of this change, indeed the only one observed with certainty, corresponds to a movement of the pole of rotation, that is a variation of latitude. An international group of stations of the *International Latitude Service* makes regular observations to measure this variation. Changes of tilt are much more local and require very sensitive tilt meters to detect. This again is a subject outside the scope of geodesy, properly speaking.

The aims of geodesy are to provide a geometrical framework on the surface of the Earth on which maps may be developed and into which other scientific data may be located, and to determine the external gravitational field. Originally the necessary observations were confined to the surface of the Earth and so, restricted by geometry and atmospheric refraction, the results were necessarily local, partial and inexact. Artificial satellites, space probes and radar measurements of distances to the Moon and planets have extended the scope of geodetic observations, liberated geodesy from its restriction to the surface, made it a potentially world-wide study and have already increased the accuracy of data. The wider scope, greater detail and refined results enable smaller features of the form and potential of the Earth to be established and a deeper insight to be obtained into the structure of the interior of the Earth, so that geodesy is now not solely providing a framework but is contributing to geophysical knowledge in a way not hitherto possible.

Further reading

No detailed references have been given to particular items of this chapter; the following works will enable individual topics to be pursued in detail.

Classical Geodesy:

Bomford, G., *Geodesy,* Oxford University Press, 1952.

Electromagnetic methods for the measurement of distance:

Electromagnetic Distance Measurement, *Proceedings of a symposium of the International Association of Geodesy,* held in Oxford, September 1965. London: Hilger & Watts, 1967.

Artificial satellites:

The use of artificial satellites for geodesy, *Proceedings of the first international symposium on the use of artificial satellites for geodesy,* ed. G. Veis., North Holland Publishing Company, Amsterdam, 1963.

Mueller, I. I., *Introduction to satellite geodesy,* Ungar, New York, 1964.

Cook, A. H., The contribution of observations of satellites to the determination of the Earth's gravitation potential space. *Space Sci. Rev.* **2** (1963) 355–437.

Fundamental Constants:

The System of Astronomical Constants, *I.A.U. Symposium No. 21*, Ed. J. Kovalevsky, Gautier Villars, Paris, 1965.

Seismology

H. M. Iyer, Geophysicist, U.S. Geological Survey, National Center for Earthquake Research, California

3.1 Seismology and Seismic Waves

Introduction

The 'terra firma' is never at rest, spinning and wobbling about on its axis, our earth races through space. Its surface is pulled and pushed by the attraction of the sun and the moon. Its inside is probably in turbulent convective motion. Volcanoes fume, shiver and erupt creating severe local tremors. Hundreds of thousands of earthquakes occur annually, few of them very severe and destructive, in the solid outer *crust* and the upper *mantle*. Every storm in the sea, with its associated wind field, atmospheric turbulance, and ocean waves cause continuous pulsations of the earth's surface. Finally there is man himself, the ultimate source of *seismic noise*, with his restless activities, the factories, the traffic and an endless array of vibration sources. Added to these, man now possesses power to simulate even large earthquakes by exploding nuclear bombs. Seismology is the measurement, analysis and study of all the movements of the solid earth.

Among the disturbances listed above, the most violent and destructive, and hence most noticeable, are earthquakes. In fact seismology began as a science of earthquakes ('seism' means shock). However, at present we study every type of earth movement, from the large earthquake waves to the minute omnipresent seismic pulsations which move the earth's crust by only a few milli-microns (one milli-micron = one millionth of a millimeter); and from very fast or 'high frequency' vibrations to very slow movements one cycle of which can take hours to complete. Thus, in seismology, we deal with a spectrum of seismic movements, like the familiar

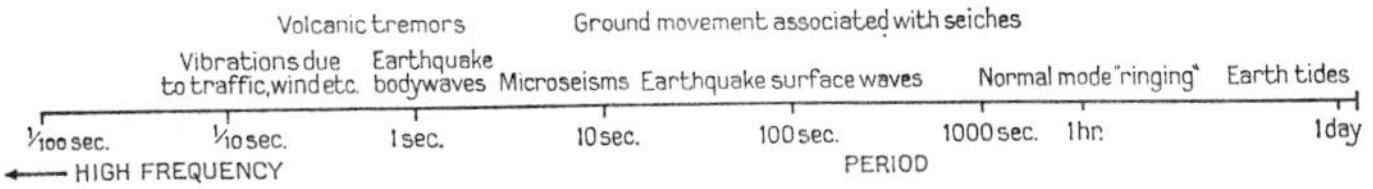

FIG 15 Seismic spectrum. (Reproduced by permission of *New Scientist*)

spectrum of optics. The shape of the spectrum is a function of time and of the place where the measurements are taken. Such a seismic spectrum, indicating the various types of earth movements, is shown in Figure 15. Later on we will see how the seismic disturbance propagates waves in the elastic earth, and how these waves travel with widely different speeds and are subject to reflection, refraction, interference, diffraction and scattering. The seismologist's job is to record seismic waves and sort them out according to their frequencies, velocities, paths of propagation, and infer from these the nature of the mechanisms that generate the waves and the elastic properties of the media through which they travel. We owe to these studies the bulk of our current knowledge about the interior of the earth and the mechanism of earthquakes.

Historical

Aristotle believed that earthquakes were caused by the escape of air within the earth. In China, as early as A.D. 143, there existed an instrument to indicate that an earthquake had occurred and to measure the direction of ground motion associated with it. In 1859, the seismological pioneer Robert Maleet studied the damage to some Appennine hilltowns caused by an earthquake. He worked a rough system to estimate the direction from which the destructive waves came, by studying the nature of ground crack and direction of fall of buildings. He also made some pioneering efforts to catalogue and understand the geographic distribution of earthquakes.

John Milne, an English engineer, can be called the founder of scientific seismology. He went to Japan in 1875, and his interest in earthquakes was aroused as he was exposed to several of them. He was directly responsible for evolving seismic recording as a precise science, and for starting co-operative international seismic efforts on a world-wide basis. During Milne's time, about fifty Milne-Shaw seismograph stations were set up in several parts of the world.

From then on, the science of seismology has progressed with rapidity, and today there are nearly 1,000 seismic stations, spread all over the world, continuously monitoring the several hundred thousand earthquakes that occur annually. The vast amount of data collected from these stations is studied and interpreted to give a detailed picture of the mechanism of earthquakes and the structure of the earth. One of the outstanding features of seismological studies is that, as in meteorology, there are no international barriers for exchange of seismic data, results of studies, and know-how. The experimental studies are supported by substantial theoretical work covering many areas and crossing the frontiers of physics, mathematics and statistics. A very practical outcome of earthquake study has been the employment of parallel techniques in oil-exploration, where shock waves from an artificial explosion are used to study the layering of the upper crust of the earth and to look for favourable oil-bearing features. Since 1959 seismology has developed at an explosive rate, backed up by the political necessity to distinguish between earthquakes and man-made nuclear explosions by seismic measurements.

Thus, today, seismology deals with the following:

(1) The practical problem of understanding, evading and living with earthquakes.
(2) The use of earthquakes and other natural excitations of the Earth to understand the nature of the terrestrial forces involved and the structure of the Earth.
(3) The technology of oil exploration.
(4) Investigations into the differences in the nature of seismic waves generated by earthquakes and underground nuclear explosions with the aim of developing a system to detect clandestine nuclear bomb tests.

Elasticity of the Earth

The apparently hard Earth is an elastic body. When a force or a 'stress' is applied, it is deformed or 'strained'. If the applied force is not large enough to cause rupture, its removal causes the earth to return to its original shape after executing an oscillatory motion. The oscillations of the excited part of the earth are transmitted to adjacent particles resulting in a continuous propagation of the disturbance, like ripples in a pool of water, the only difference being that, in the earth, these ripples can travel in all three dimensions at great speeds. The frequency and amplitude of the oscillations at the origin depend on the stress applied and the elastic properties of the material in the vicinity. Because the Earth is made up of material with elastic parameters varying along the surface and with depth, the 'ripples' are modified and distorted, the nearest two-dimensional analogy being a shallow choppy sea with lots of islands.

Body waves

One of the basic theorems of elasticity is that an elastic medium can be subjected to two types of deformation: compression and shear. Hence all the elastic waves detected in seismology are basically *compressional* or *shear* waves. In compressional waves, particles *push* one another and the waves propagate as alternative compressions and rarefactions along the direction of travel, as in sound waves. The shear waves propagate by particles brushing against one another by rotational movements, at right angles to the direction of travel. The closest analogies to this type of wave are the vibrations of strings and light waves. The wave motion in shear waves is termed rotational or equivoluminal.

The compressional waves of seismology are generally termed *P-waves*, short for 'Primary' or 'Push' waves. The shear waves are called *S-waves*, short for 'secondary' or 'shake' waves. Both together are referred to as *body waves* because they tend to cut across the interior of the Earth. Figure 16 shows the particle motions in longitudinal and transverse seismic waves.

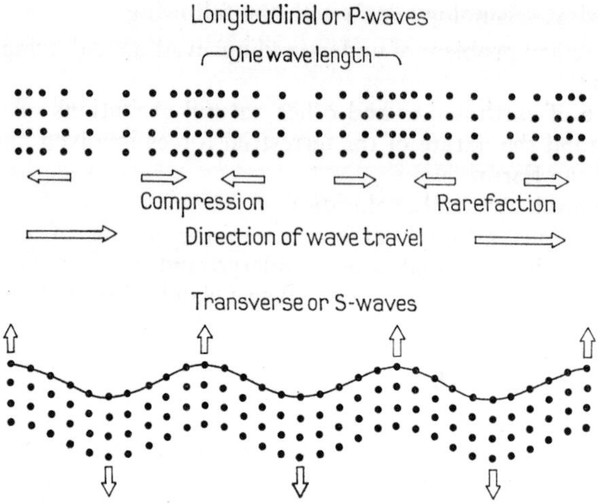

FIG 16 Particle movements in longitudinal and transverse seismic waves. (Courtesy of Harper and Row, Publishers)

The speeds of travel of *P*-waves and *S*-waves are determined by the two elastic parameters called Lamé Constants, pertaining to the medium of travel. These are commonly denoted by the Greek letters λ and μ. Of the two Lamé Constants, μ has a relatively simple physical significance. It describes the resistance of the elastic material to a shear type of deformation, and is generally known as the *shear modulus of elasticity*. The constant λ, on the other hand, is related both to the shear modulus μ, and the *bulk modulus of elasticity*, k, by the following equation:

$$\lambda = k - \tfrac{2}{3}\mu$$

The velocities of *P*- and *S*-waves are given by the following equations involving λ, μ and the density, ρ.

$$P\text{-wave velocity,} \qquad \alpha = \sqrt{\frac{\lambda + 2\mu}{\rho}}$$

$$S\text{-wave velocity,} \qquad \beta = \sqrt{\mu/\rho}$$

For many solids (in particular, for the rocks which make up the bulk of the earth) it is found that the two Lamé Constants are the same. Using this relation, it can easily be seen that, for such material, $\alpha = \sqrt{3}\beta$. Thus from a seismic event, the *P*-wave travels about 1·7 times faster than the *S*-wave.

Surface waves

Lord Rayleigh first proved in 1885 that, in addition to body waves, a type of wave could travel along the Earth's surface. These waves are referred to as *Rayleigh waves* and can be compared to waves on the surface of the ocean. The Rayleigh waves were first identified on an earthquake record (called *seismogram*), in 1889. In Rayleigh waves the Earth particles move in an ellipse, counter-clockwise when the waves moves from left to right. There is no movement at right angles to the direction of wave travel, the motion being entirely confined to the vertical plane passing through the direction of propagation. The Rayleigh wave velocity is about 0·9 times the velocity of *S*-waves.

The body wave and Rayleigh wave descriptions outlined so far assume a simple model for the earth's structure. However, in reality, the Earth has rather a complex structure, the density and elasticity varying continuously, and sometimes quite sharply, between the surface and centre of the Earth. These changes complicate the simple pattern of body and surface waves, and in practice one has to deal with a mixture of reflected, refracted and guided waves.

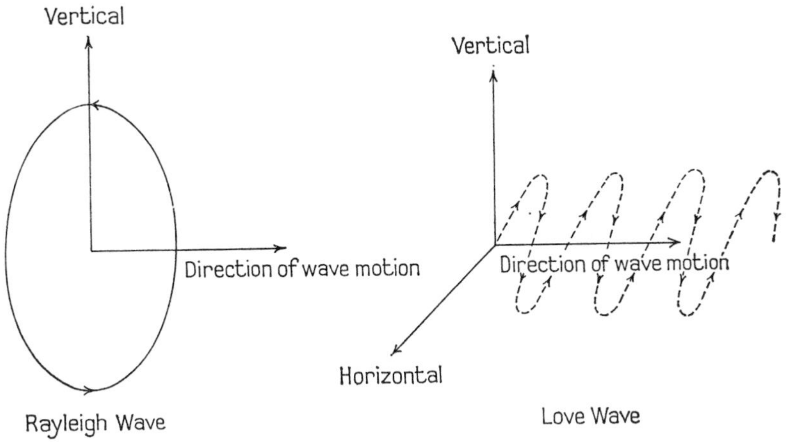

FIG 17 Particle movements in Rayleigh and Love waves

A first consequence of this deviation from homogeneity is the generation of *Love waves*. In studying the large, long-period Rayleigh waves, very clearly identified in seismograms, it was found that not all the particle motion was of the retrograde-ellipse-in-a-vertical-plane type. There were strong indications to show that transverse motion in a horizontal plane was definitely present in the recorded Rayleigh waves. A. E. H. Love, a leading British authority on the theory of elasticity, investigated this phenomenon and established that these waves were a type of guided waves, caused by multiple reflections of body waves between the surface of the Earth and some boundary below, where the elastic properties changed sharply. In 1909, Mohorovičić discovered, from *P*-wave studies, that the elastic discontinuity required for such wave reflections did exist, at a depth of about 30–40 km under continents and about 10 km under oceans. The transverse waves are known as *Love waves*. The surface displacements in them are entirely transverse and in the horizontal plane. The Love wave velocity is the same as that for *S*-waves under the discontinuity discovered by Mohorovičić.

The Rayleigh and Love Waves together are generally referred to as *surface waves*. The types of particle motion in Rayleigh and Love waves are schematically illustrated in Figure 17.

3.2 Seismological Instruments

Principles of seismometry

The instrumentation system used to detect earth movement and record it in an understandable form is called a 'seismograph'. It consists of a 'seismometer,' which is the mechanical part reponding to the movement, a mechanical or electrical amplifying device, and a recording part.

As was mentioned earlier, a crude seismometer did exist in China during the second century A.D. It consisted of a ring of eight dragons holding balls loosely in their mouths. When the apparatus was shaken, the balls fell from the mouth of the dragons facing the direction of ground motion. Another early device consisted of a bowl, with a ring of spouts, in which mercury was kept. When the bowl was shaken, mercury overflowed through the spouts. A more sophisticated device was used by Palmeri in Italy, in 1866. This device made an electrical contact when an earthquake occurred, which stopped a clock and at the same time started a recording drum. A pencil was made to draw a line on the drum as long as the vibrations lasted. Thus Palmeri's seismograph was able to give the time of occurrence and duration of motion in earthquakes.

The seismometer utilizes the property of mass called inertia. The mechanical part consists of a free moving pendulum, usually referred to as a 'boom', attached to a frame rigidly coupled to earth. When the frame moves with the Earth, the pendulum lags behind because of its inertia, and thus a relative motion occurs between the boom and the frame. The situation is similar to the forward motion of passengers in a bus when the driver applies the brakes sharply. The pendulum has its own natural frequency. When the earth movements are of very high frequency compared to this natural frequency, the pendulum is unable to follow the movements and remains stationary in space. In this situation, there is maximum relative movement between the frame and the boom. Thus at very high frequencies, compared to its natural frequency, the seismometer response is said to be '1'. At low frequencies, however, the pendulum is able to follow the frame movements faithfully. Hence there is no relative motion and the response of the seismometer is said to be '0', at very low frequencies. In between there is the familiar resonance frequency when the earth movement frequency coincides with the seismometer natural frequency, and the response can reach a theoretical maximum of infinity.

The response of a seismometer is defined as the ratio of the movement of the tip of the boom to the actual earth movement. In practice, the response of a seismometer is not only determined by its natural frequency, but also by the damping factor, that is, the resistance to pendulum movement. Seismometers are usually operated at 'critical damping', which means that when the boom is disturbed from its normal rest-position, it returns there without any overshoot. Critical damping is generally achieved by using oil in a dash pot or by generating eddy currents in a copper conductor or coil attached to the boom and held between the poles of a magnet fixed to the frame.

As in any three-dimensional phenomenon, to define the complete earth movement requires it to be measured along x, y and z axes. Hence a complete seismometer set must consist of three instruments, two horizontals at right angles to each other and one vertical. Usually the two horizontals are kept in such a way that the freedom of movement of the pendulums are in the north-south and east-west directions, respectively. Figure 18 shows diagrams of simple horizontal and vertical seismometers.

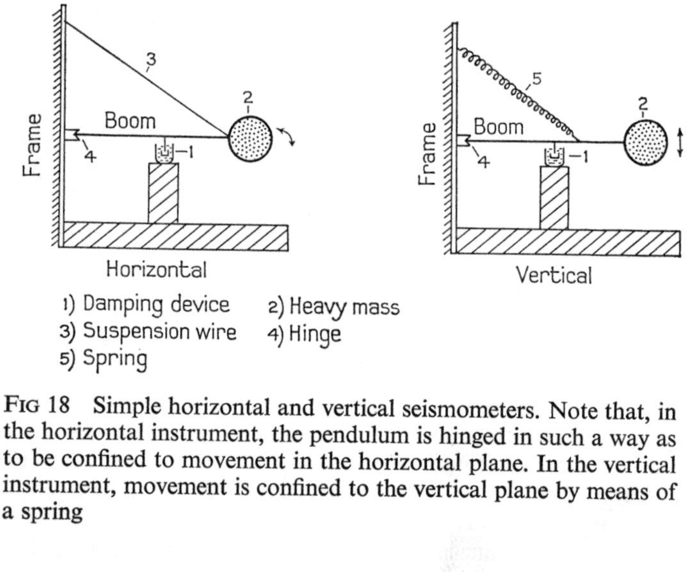

FIG 18 Simple horizontal and vertical seismometers. Note that, in the horizontal instrument, the pendulum is hinged in such a way as to be confined to movement in the horizontal plane. In the vertical instrument, movement is confined to the vertical plane by means of a spring

Modern developments

Once the relative earth motion has been detected by a seismometer, the next step is to amplify and record the information. In the earliest instruments, this amplification of the boom movement was achieved by using a system of mechanical levers and recorded by a stylus scratching on smoked paper. One such classic example is the *Wiechert Seismograph* where a mass of the order of 17 tons is supported on a vertical column, which acted as the boom.

The next innovation was the use of an optical lever to magnify seismometer movement. A light beam originating from a lamp is reflected by a mirror on the boom, and recorded photographically. Seismographs using this principle are the Milne-Shaw and Wood-Anderson instruments.

In 1914, Galitizin introduced a great innovation in seismographs by using an electromagnetic transducer on a seismometer to convert earth motion into an electrical signal. The transducer consisted of a coil of copper wire on the boom which moved in the gap of a permanent magnet attached to the frame. Galitizin fed the electrical output from the transducer to a sensitive moving-coil spot-light mirror galvanometer. The movements of the light spot could be recorded photographically. The introduction of the galvanometer adds its own natural frequency and damping factor in the final response equation of the total system. In addition, the electro-magnetic characteristics of the seismometer transducer (as well as that of the moving coil galvanometer) enter as parameters in the equation. By a proper choice of the six variables, two natural frequencies, two damping factors, two coil-magnet parameters, the final shape of the response curve as a function of frequency can be varied over a wide limit. Because of this flexibility and simplicity the Galitizin system has found unchallenged acceptance and occupies a prominent place even today. It has been found that for earth frequencies in the neighbourhood of 1 Hertz (1 c.p.s.), the earth movement can be amplified by a factor of a million. Thus, deflection of the light spot by a millimeter means that the earth has moved by only as much as one two thousandth of the wave length of visible light!

With the rapid progress in electronics, special amplifiers have been developed to magnify the minute low frequency electric signals—much lower than those conventionally encountered in audio work—that are detected by Galitizin seismometers. In recent years, very compact, high magnification, transistorized, low-frequency amplifiers have become commercially available.

The use of smoked paper and photographic paper for obtaining the amplified record of earth motion has already been mentioned. When electronic amplifiers are used, the output from them can be recorded using a pen recorder, conventionally referred to as a strip-chart recorder. Whatever may be the final form of recording, since the instruments have to be monitoring events continuously, it is not desirable to use a long strip of paper as prodigious amounts of it will be consumed every day. Hence seismologists have preferred to get a whole day's record on a single rectangular sheet of paper of reasonable size. A helical recorder is used for this purpose. In this recorder, the sheet of paper is mounted on a drum, which rotates at a constant speed, the position of the pen changing slowly in relation to the width of the paper.

An essential part of the recording system is an accurate timer. Time marks are provided on the records by deflecting the trace for a short interval of time or, with photographic recorders, by turning off the light for a second or two. Usually the time marks are made at the end of each minute, a longer time-mark indicating the end of the hour. The timer normally runs for many days without appreciable error, and its accuracy is checked by comparison with radio time signals. (In modern stations the timers are crystal clocks accurate to within a second in a month). A typical seismic record taken on a helical recorder shows minute and hour marks, and also continuous wiggly lines. These indicate *microseisms*, the ever present background seismic noise of the earth.

The selection of seismographs for a particular job depends on the frequency characteristics of the seismic waves to be studied. For studies of events like rock bursts in mines, explosions, local earthquakes within say 10–20 km of the recording station, the most important information is contained in frequencies in the region of 5–10 Hz, and a magnification of about 100 000 is suitable. At greater distances, the *P*-waves are better recorded by instruments with natural frequency near 1 Hz and magnification of 10 000 to 100 000. For *S*-wave studies, lower natural frequency instruments are desirable. Low-frequency seismographs with a natural period of about 30 seconds (0·033 Hz natural frequency) are used for recording surface waves.

Since the early days of the Wiechert and Milne-Shaw instruments, seismographs are becoming progressively better engineered, smaller, more streamlined and more sensitive.

The great bulk of our seismological studies are based on 'reading' the seismograms produced by simple helical recorders. However, progress demands special high resolution studies and the conventional record does not satisfy these requirements. Some improvement in quality of record and a certain amount of flexibility can be obtained by using a strip-chart recorder. Better still is the use of tape recorders. As the seismic signals lie in the sub-audio range, conventional tape recorders are of no use. However, the technology of recording ultra low-frequency signals on magnetic tape, using the principle of frequency modulation, has reached a very high degree of development in recent years, and very compact, reliable tape recorders using transistors and precision components are now commercially available. A great advantage of tape recorders is, of course, that the recorded tape can be played back much faster than the recording speed. In addition, the output is in the form of electrical signals and can be handled by other analysing instruments.

In these days of fast digital computers, any data needing large-scale collection and processing should be in a form that can be directly fed to the computers. The food for computers is in the form of numbers, and the instruments that convert continuously varying electrical signals into discrete numbers are called 'analogue to digital converters' or simply *digitizers*. Digitizers can handle directly data flowing out of a seismic system in what is called 'on line' operation. Alternatively, data recorded on magnetic tape can be played back 'off line' and digitized. The use of digitization and computer processing of data has created a minor revolution in seismology. Many complex operations, once beyond dreams, are now being done using computers that can look through a whole day's data in a matter of few minutes. One could hopefully extrapolate and predict that, in the not-too-distant future, a world-wide network of stations may be continuously transmitting seismic data, via satellite communication systems, to a central 'international' computer, programmed to tell with little time delay when and where an earthquake has occurred.

Seismological observations

Since Milne's days, the importance of continuously recording seismic observatories, operated at several points on the Earth's surface, has been recognized. The requirements of a good seismic observatory are:

(1) It should have a number of horizontal and vertical instruments capable of good coverage over the seismic spectrum. In practice six instruments, one short period three component set and one long-period three component set, are adequate.
(2) It should have a good recording and timing system.
(3) It should be located as far away as possible from artificial seismic noise sources like railways, quarries, mines, factories. In general it is not advisable to have a seismic observatory located in the heart of a big city.
(4) As the instruments are very susceptible to spurious pick-ups from convection currents in the instrument room, the temperature should be kept fairly steady, and the room should be free from draughts and sudden pressure fluctuations.
(5) It is desirable to keep the seismometers in a special 'seismic vault', preferably underground, to help in maintaining a steady temperature. The seismometers are placed on heavy concrete plynths in firm contact with bed rock.

The design and response characteristics of seismographs operated in different parts of the world vary considerably. The difficulty of making global studies with such non-standard data has set severe limitations in seismology. The age-long dream of seismologists has been to have a world-wide network of seismographs with identical response characteristics and standardized recording and timing systems. Such a dream is coming true with the active progress being made by the U.S. Coast and Geodetic Survey in setting up 125 standard seismic stations all over the world (see Figure 37). There is a World-wide Standard Seismic Station, where the seismic vault is designed according to the rigid specifications outlined in this section, in operation at Eskdalemuir, Scotland.

Special purpose instruments

Displacement meters

The coil-magnet transducers produce electrical signals which are proportional to the *velocity* of the ground. As velocity is the time derivative of displacement, in simple harmonic motion at very low frequencies the velocity is very small (velocity $= 2\pi \times$ frequency $\times$ displacement) and the transducers are not very efficient. Hence displacement transducers are used in very long period seismographs. In one such instrument, an element on the boom and another on the frame together form a parallel-plate condenser, which is made part of an L–C circuit generating radio frequencies. As the spacing between the plates of the condenser varies due to earth movement, the radio-frequency changes. Thus the change in frequency is a measure of earth movement. As long period seismometers of the above type are very much susceptible to permanent displacements of the ground, convection currents etc., some form of automatic centering devices using negative feed-back is usually incorporated in them.

Accelerometers and strong-motion seismometers

For studies of very local earthquakes, and for assessing the forces produced by such earthquakes on man-made structures, the problem is one of measuring forces, that is accelerations, rather than velocities or displacements. Using the simple differential equation of the seismometer, it can be proved that the boom displacement becomes proportional to ground acceleration when the seismometer natural frequency is much larger than that of the ground frequencies being measured. Such instruments are referred to as *accelerometers* and are generally used in an overdamped condition. The use of accelerometers in engineering studies of earthquake resistant structures will be discussed in a later section.

Strain seismometers

As earthquakes are caused by the accumulation of strains within the Earth, and seismic waves themselves propagate by straining the material in their path, it is of great interest to measure such strains directly. The instruments used for this purpose are called *strainmeters*. In one strainmeter, designed by Dr. Benioff, the change caused by strain in the relative positions of two piers anchored to bedrock is measured. A rigid tube of steel or fused quartz attached to one of the piers extends to within a short distance of the other pier. A transducer introduced in the small gap between the end of the tube and the second pier gives an electrical signal proportional to gap length and hence to strain. Figure 19 shows the principle of the strain seismometer.

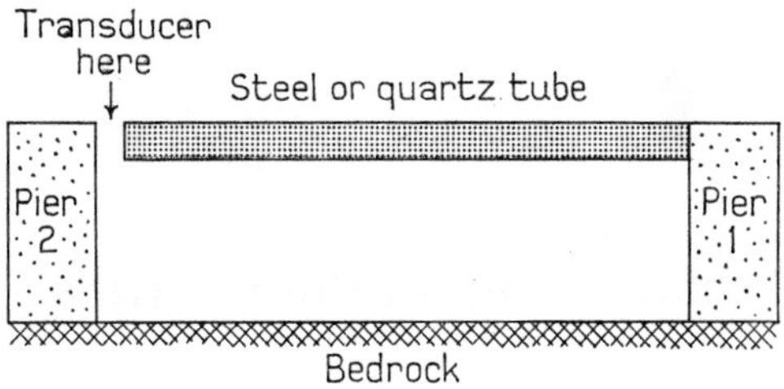

FIG 19 Schematic diagram of Benioff's strainmeter

Lunar and ocean bottom seismometers

In recent years very rugged and compact seismometers have been designed for operation on the deep ocean bottom. Seismometers with capability to telemeter data to Earth, have been designed for operation on the Moon. These instruments will be described in later sections on space and ocean bottom seismology.

3.3 The Seismogram and the Earth's Structure

The seismogram

We have seen how a disturbance within the Earth can generate seismic waves which travel at different speeds. We have also seen how seismographs can pick up the earth movements generated during the passage of such waves, amplify and record them. Figure 20 shows an actual seismogram, the record of an earthquake from a station in New Zealand.

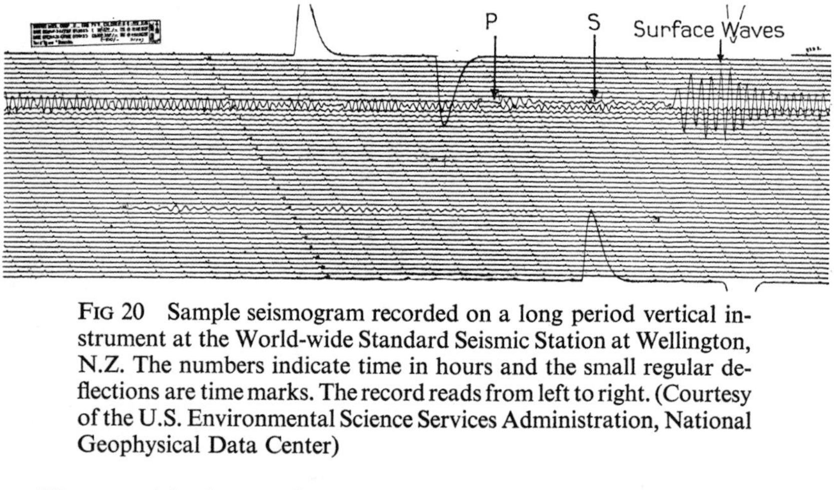

FIG 20 Sample seismogram recorded on a long period vertical instrument at the World-wide Standard Seismic Station at Wellington, N.Z. The numbers indicate time in hours and the small regular deflections are time marks. The record reads from left to right. (Courtesy of the U.S. Environmental Science Services Administration, National Geophysical Data Center)

The record is characterized by several short wave groups, clearly separated in the beginning, developing into large waves that slowly die off. The wave groups are *P*-waves and *S*-waves, which travel through the interior of the earth by different paths. The large amplitude waves of long duration are surface waves. The record is quite simple in the sense that the wave types, especially the *P* and *S*, stand out clearly. This is because it was from a long-period seismograph and also because the event was *teleseismic*, that is, at a large distance from the recording station. However, usually seismologists have to contend with very complex records and to unscramble the fund of information contained in the seismogram.

The various wave groups in a seismogram are referred to as *seismic phases*. The region inside the earth, where the earthquake occurs is called the *focus* and the place on the surface vertically above is the *epicentre*.

Refraction of seismic waves

One of the most significant early findings in seismology was that the speed of travel of seismic waves increased as they travelled deeper within the earth. This phenomenon was established by observations which showed that seismic body-wave velocity, computed by dividing the distance between epicentre and recording station (termed *epicentral distance*) increased with the distance. If it is assumed that the waves observed at larger distances had to travel through deeper regions within the Earth, the above observations mean that the *waves travelled at much higher velocities the deeper they penetrated inside the Earth.* On examining the equations for the body-wave velocities (see page 72), we note that they are of the form (elasticity/density)$^{\frac{1}{2}}$. Since density can be expected to increase within the Earth, that part alone should cause a lessening of velocity with depth. Hence, to fit with observed facts, it was logical to conclude that elasticity increased with depth towards the interior of the Earth, indicating a change in mineral composition and physical state of the rocks within.

The above discussion shows that the principle of refraction of seismic waves forms a powerful tool in seismology. The refraction of a seismic wave caused by increase in wave velocity causes it slowly to bend upwards until it ultimately returns to the earth's surface. However, within the Earth there is not only a continuous increase, with depth, in wave velocity, but also layering, with abrupt change in elastic properties across sharp boundaries. When waves strike such layers they are partly reflected and partly

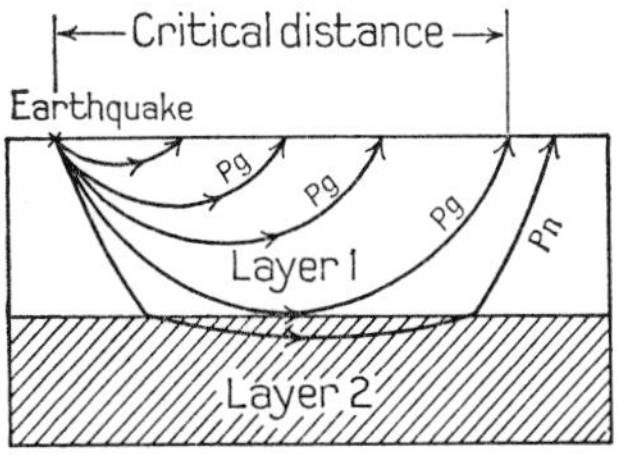

FIG 21 Refraction and reflection of seismic body waves caused by increase of velocity with depth and layering within the Earth. The seismic *phases* caused by such effects are also shown

refracted, as long as the angle of incidence is less than the critical angle. At the critical angle of incidence, the wave travels along the boundary between the two layers for some distance until it slowly turns upward to return to the surface. Figure 21 illustrates these ideas.

By studying seismograms from several observatories, the seismologist reconstructs the ray paths and wave velocities within the Earth.

Internal structure of the Earth

The paths of seismic waves are affected by two chief discontinuities within the Earth, at depths of about 30 to 60 km and 2,900 km respectively. Of these the first one, called the Mohorovičić Discontinuity, commonly referred to as the 'Moho' has already been mentioned. These two discontinuities divide the Earth into three parts, *the crust, the mantle* and *the core.*

The crust and its effect on seismic waves

Referring again to Figure 21 it will be seen that, within the critical distance prior to the ray's grazing entry into the second layer, the seismic pulses should consist of first layer refracted phases (caused by bending of the ray due to velocity increase in the first layer itself) and reflected phases (not shown). Beyond the critical distance, the first pulse to arrive is that refracted through the second layer, where velocity is much larger than at the bottom of the first layer. In fact it was the identification of the small long-period motion prior to the first layer phases in seismograms that led to the discovery of the Moho.

The first layer refracted phases are referred to as P_g and S_g, the reflected phases by repetition of letters, *PP* and *SS*, and the Moho phase is called P_n Within the critical distance, the simple crust model indicates that the order of arrival is P_g, S_g, *PP*, *SS*. Beyond the critical distance, the order of arrival is P_n, followed by the S_n and reflected phases.

The velocities of these phases are approximately as follows:

$$P_g = 6 \text{ km/sec}$$
$$S_g = 3{\cdot}5 \text{ km/sec}$$
$$P_n = 8{\cdot}1 \text{ km/sec}$$
$$S_n = 4{\cdot}4 \text{ km/sec}$$

Later studies have shown that, besides these simple phases, there is a host of reflected and refracted arrivals, indicating the presence of several more layers within the crust itself.

The Earth's core and the 'shadow zone'

When seismograms from successively more distant observatories are examined for a single large earthquake, a sharp change in the *P* and *S* pattern becomes apparent beyond 11 000 km from the epicentre. Up to that distance the first arrival is the *P*-wave travelling through the mantle, its velocity increasing smoothly with distance. But beyond 11 000 km the *P*-pulse is not seen at the expected time of arrival corresponding to this smooth variation. What is worse, the *S*-pulse disappears altogether! From

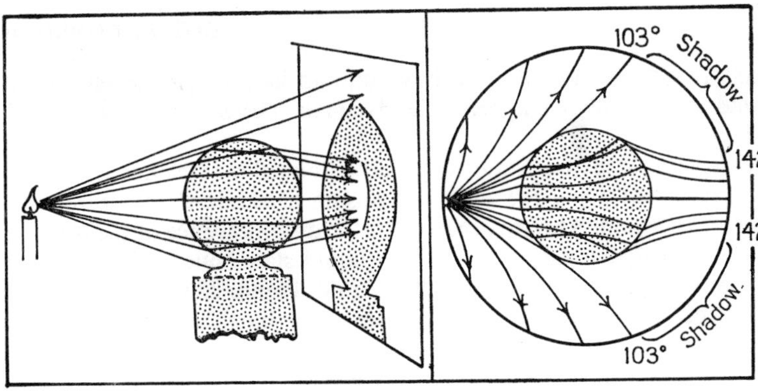

(a)

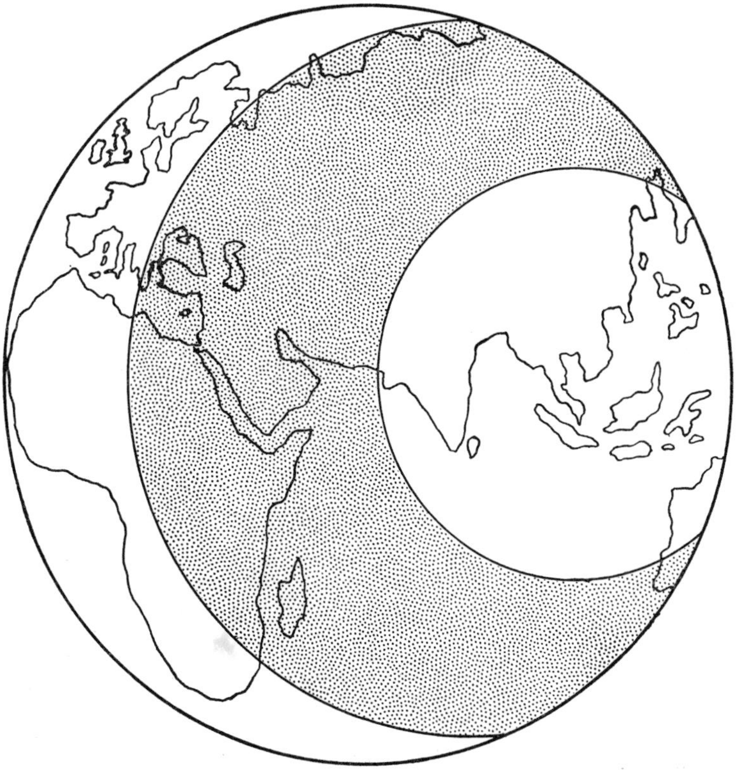

(b)

FIG 22 (a) How the Earth's core causes a shadow zone.
(b) Extent of shadow zone for an earthquake at Peru. The sharpness of the zone limits is exaggerated. (From *Biography of the Earth* by George Gamow. Copyright © renewed 1969. By permission of The Viking Press, Inc.

11 000–15 000 km, the *P*-pulse is weak, and its arrival times erratic, but beyond 15 000 km, it re-appears clearly again. The *S*-wave is not seen after 11 000 km.

To explain these observations, it was postulated that the Earth had a core which refracted away the *P*-wave, causing a 'shadow' in the region of 11 000–15 000 km. Also, to account for the disappearance of the *S*-wave, it was concluded that the core was in a liquid state, since it is known that shear waves cannot be sent through a liquid. Figure 22 shows the mechanism of occurrence of a *shadow zone* due to the Earth's core. The extent of shadow zone on the earth's surface for an earthquake is also shown.

Before proceeding further it will be useful to define distance measurements in degrees rather than miles or kms. Because of the spherical shape of the Earth, such angular units are much more convenient to use in seismology. The distance, in degrees, between any two points on the surface of the globe is the angle subtended by the arc joining the points to the Earth's centre. It may be noted that one degree is equivalent to very nearly 111 km on the linear scale. Thus the core shadow zone covers the region 103°–142°.

The principal refracted phases and their paths due to propagation in the mantle and the core, and the formation of the shadow zone, are shown in Figure 23. The phases refracted through the core are called *PKP* (or *P'*). Note that, within the shadow zone, only low amplitude diffracted *P* and *S* waves are normally possible. However, it was observed that, even at dis-

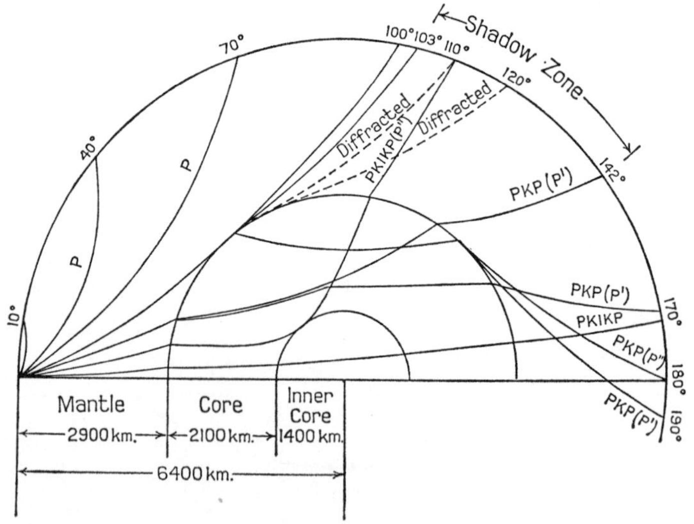

FIG 23 The principal refracted phases due to propagation in the mantle and core of the Earth. Both compressional and shear phases exist for every path, except when it involves travel through the liquid core, when shear waves vanish

tances as low as 110°, there were strong arrivals on seismograms. This observation was quite puzzling at a time as geometry of the core was fairly well established. The puzzle was solved by the Danish seismologist, Miss Lehman, by postulating the existence of an inner core at a depth of nearly 5,000 kms from the surface. The inner core phase is designated as *PKIKP* (*P″*). Both compressional and shear phases exist for every path except when it involves travel through the liquid core, when shear waves vanish.

It should be remembered that, when we talk of specific phases, the aim is to indicate their importance in our understanding the structure of the Earth. This does not mean that an earthquake record shows only a few specific phases, depending on the epicentral distance, and blank everywhere else. In practice several scores of combinations of *P*- and *S*-waves, bouncing around inside the earth, make up the seismogram. The larger the earthquake, the longer the reverberations last.

Reflected seismic waves

Reflection phases at discontinuities like the Moho, mantle-core boundary, and several other less defined boundaries, are indicated by repetition of letters like *PP*, *SS*, etc. Since *P* can be transferred to *S* and vice versa during the process of reflection, the reader can well imagine the variety of possibilities.

Surface reflections can cause phases such as *PP*, *SS*, *PS*, *SP*, *PPP*, *SSS*, etc.

Core reflections cause *PcP*, *ScS*, *PcS*, *ScP* phases.

Reflection and refraction combined cause phases such as *PKKP*, *PKKS*, *SKKP*, *SKKS*, etc. (Note that the symbol *K* indicates travel through the

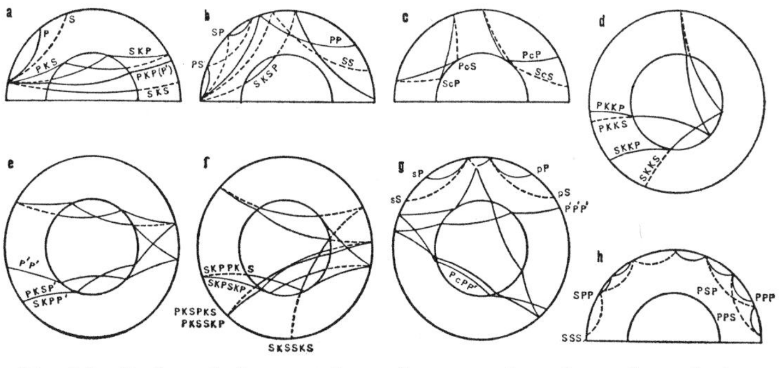

FIG 24 Reflected phases at the surface, mantle and core boundaries. Longitudinal wave shown as full lines, transverse waves shown dashed. (From *Elementary Seismology* by Charles F. Richter. W. H. Freeman & Co. Copyright © 1958)

core as compressional wave.) In Figure 24 several possible reflected and refracted phases of waves with letter symbols are shown.

The lower case letters *p* and *s* are used to describe a special type of reflection, which will be discussed under deep-focus earthquakes.

Surface waves

The surface waves constitute the most prominent part of a seismogram. They are characterized by large, long-period, regular looking waves containing, besides the Rayleigh and Love phases already described, a wide variety of guided waves. The two important properties of surface waves are their propagation in two dimensions and *dispersion*.

Dispersion is a property of waves, by which the velocity of travel depends in some way on the frequency. The seismic disturbance, which is a short pulse at the source, can be treated as an aggregate of simple harmonic waves covering a wide range of frequencies. Each of the elementary-frequency waves train travels with a different speed. Thus the pulse is spread out into a long train of sinusoidal oscillations. The dispersive property of surface waves is directly dependent on the structure of the material of the earth through which they travel. The longer the period of the wave, the larger its wave length and the fewer the discontinuities it 'sees' within the Earth. As the periods get shorter, the wave is affected by finer structure of the Earth. In fact this is the cause of the dispersion observed. The phenomenon is analogous to travelling in a tractor, truck, car and scooter, over a poor road. The tractor, with its large tyres, does not 'see' the road roughness at all; the truck sees the larger ones, the car bumps a little more, and the scooter bounces even over the small stones. Thus the tyre size, in the example, can be compared to wavelength.

Surface wave dispersion study forms a very powerful tool in understanding the earth's structure. A substantial amount of our knowledge of the large-scale structure of the earth's crust and upper mantle, and the difference between continental and oceanic crusts, has been derived from the study of surface-wave dispersion. The technique employed is to read the wave periods on a seismogram as time advances and, knowing the epicentre and origin time of the earthquake, to convert this information into a velocity-period curve, or a *dispersion curve*. This is then compared with theoretical curves based on models of the Earth's structure. The best fitting model is accepted. Theoretical computation of model dispersion curves is an advanced and complicated mathematical operation and has become possible only with the advent of fast digital computers.

The structure of Earth—a summary

Crustal structure

Mohorovičić's early estimation of the average crustal thickness was 50 km. However, a large quantity of data from body wave, surface wave, and explosion studies have indicated that, in continental regions, the total crustal thickness is of the order of 30 or 40 km. There is also some evidence that, in certain continental areas, the crust is composed of more than one well defined layer. Under mountain ranges, like the Alps and Sierra Nevada, there is marked increase in the depth of the Mohorovičić discontinuity. Under oceans, on the other hand, it has been found that the crustal thickness is less than 10 km below the ocean floor. Apart from the mountain, ocean and continental systems, the crustal thickness is atypical along the margins between continents and oceans, and under oceanic ridges, and trenches. At continental margins the crust layers varies from the 30–40 km 'continental' to the 10 km 'oceanic' structure. Very erratic depths for the Moho have been found under the mid-Atlantic ridge. Figure 25 is a schematic representation showing the layers of the earth's crust and upper mantle under continents and ocean basins. Corresponding *P*-wave velocities and geological structure (speculative) are also shown.

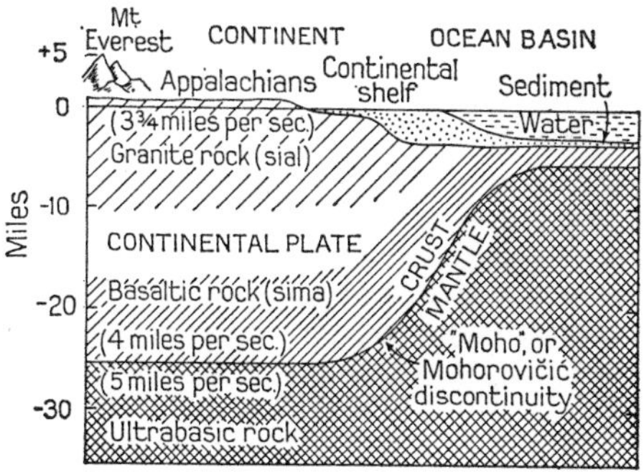

FIG 25 Layering of the Earth's crust and upper mantle under Continents and ocean basins. (Courtesy of John Wiley & Sons Inc)

What is the physical change in rocks that cause the Moho? Geophysicists have been trying to answer this question using seismological evidence, geological studies and laboratory experiments in which rocks are subjected to temperature and pressure conditions similar to those likely to be found inside the earth. The quantities that can be measured are density and elasticity from body wave velocities. These parameters are then fitted to possible rock compositions under pressure and heating. The real nature of the material is a matter for considerable speculation. The direct answer will be found only by drilling a hole through the Earth's crust and bringing up a sample of the material underneath. This was a science fiction idea as old as Jules Verne, but in twentieth century it may become a reality. The Project Mohole contemplated by the U.S.A., proposed to drill such a hole through the earth's crust under the Pacific Ocean.

The Earth's interior

Within the mantle there are several levels, near which there is departure from the regular increase of velocity with depth. Based on observations of break in traveltime curves at distances near 20° (the so called 20° discontinuity) Jeffreys postulated the existence of a marked rate of increase in seismic velocity with depth near 480 km. This transition is now well established and is considered to be caused by chemical phase change of the mantle material under high temperatures and pressures. Another rapid increase in seismic velocity has been observed at about 600 km under North America and Japan. Based on travel-time and amplitude studies of seismic data in California and elsewhere, Gutenberg proposed that immediately below the Moho there could be a zone in which the seismic velocities were lower than above or below. Such low velocity channels have since been proposed to account for seismic observations in different parts of the earth. Surface wave studies show that, under ocean basins, there is some certainty that shear wave low velocity zones exist. One possible cause for the low velocity is partial melting of rocks under temperature and pressure effects. If this is true, low velocity layers take a very important place in modern geophysics especially in the light of the new sea floor spreading theory (see later sec-

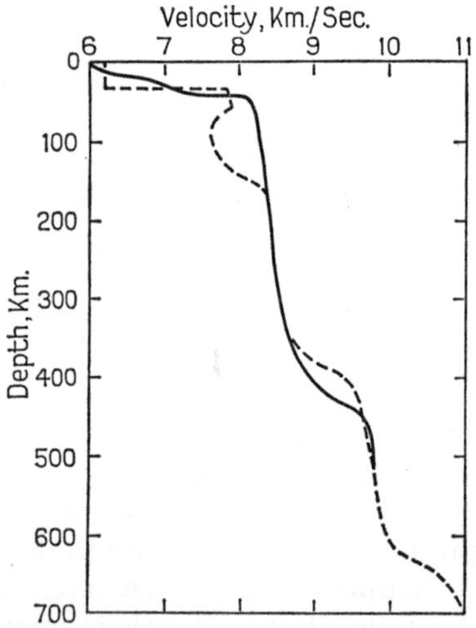

FIG 26 (a) A comparison of the *P*-wave velocity structure in the upper mantle. The solid line represents the average velocity structure for the whole of the North American continent (Iyer, *et al*, *Journal of Geophysical Research,* Vol. 74, No. 17, August 15, 1969). The dashed line is the structure under the tectonically active western part of North America (Johnson, *Journal of Geophysical Research,* Vol. 72, No. 24, December 15, 1967).

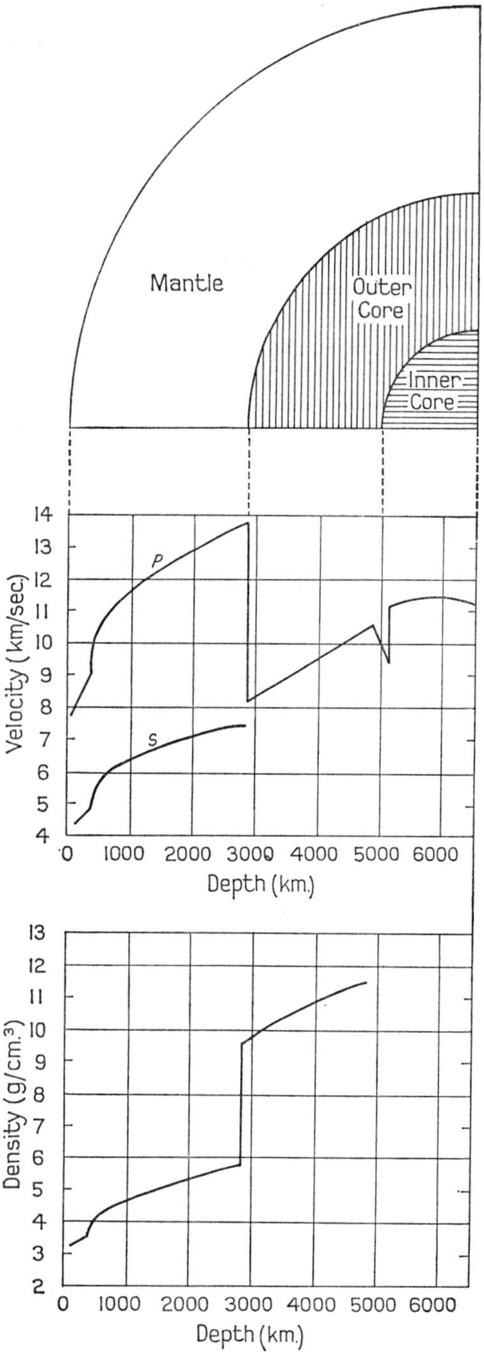

FIG 26 (b) Diagrammatic representation of the variation of *P*- and *S*-velocities and density with depth within the Earth. (From *An Introduction to the Theory of Seismology* by K. E. Bullen. Courtesy of Cambridge University Press)

tion). It has now been fairly well established that the top 50 to 100 km of the sea floor is moving at the rate of a few centimetres per year, like a giant conveyor belt. The partially molten material of the low velocity layer could provide an ideal medium for the sea floor to glide over. Unfortunately there is as yet no positive evidence to establish the existence of low velocity layers in the upper mantle. On the other hand, recent work in the United States makes it fairly certain that there is no low velocity layer under the stable part (central and eastern part) of the north American continent. Lane Johnson has shown that there is a low velocity layer under the Basin and Range province in the tectonically active Western United States. Figure 26(a) shows a comparison of the *P*-wave velocity structure in the upper mantle for the tectonically active western side (Lane Johnson) and stable central and eastern side of North America (Iyer et al.).

The mantle-core boundary appears to be at a nearly constant depth of about 2,900 km, and the transition from the outer to the inner core is at a depth of about 5,000 km. Figure 26(b) shows the distribution of velocities and density with depth throughout the Earth (after Jeffreys, Gutenberg, and Bullen).

It seems reasonable to assume that the principal constituent of the upper mantle rocks is olivine ($(Mg,Fe)_2\ SiO_4$). The velocity transitions at 400 and 600 km depths can be attributed to phase change of olivine to denser forms like spinel. Prior to 1948 it was believed that the core consisted of iron or nickel iron. Recent studies, however, show that the core may also have silicone and hydrogen in its makeup.

3.4 Quantitative Seismology

Travel-time curves

The seismogram contains the basic information on when and where an earthquake occurred and how strong it was. The information has to be derived from the times of arrival of several phases and their amplitudes.

If a particular seismic phase can be identified at several seismic stations set out on the surface of the Earth, and the time taken to reach the stations is plotted as a function of distance from the epicentre, the result is a *travel-time curve.* Since the phase travels through a particular layer and has an almost constant velocity, the travel-time curve will approximate to a straight line. When several phases are identified and their travel-time curves plotted, one gets a system of straight lines; and, since velocity = distance/time, the slope of each straight line gives the velocity of the corresponding phase. The principle of the travel-time curve is illustrated in Figure 27. Ray paths are shown for a simple two layer model with average velocities of 6 km/sec and 8 km/sec respectively. The two segments of the travel-time curve represent ray paths in the upper medium and the combined layers. The slopes of the travel-time curve and hence the velocities are determined by the layer which predominantly controls the ray paths.

In the above discussion it is assumed that the epicentral distance and origin time of the event is known. However, this was not so at the beginning of seismology. The basic data used in obtaining earthquake source information are the velocity differences between phases. Taking the simple instance of a *P* and *S*, if *x* is the actual distance along the Earth's surface from the source to the recording station, the travel times can be written as,

$$t_p = \frac{x}{\alpha}$$

$$t_s = \frac{x}{\beta}$$

If $(t_p - t_s)$ can be evaluated, then, knowing α and β, x can be computed. Obviously, $(t_p - t_s)$ will also depend on factors like the depth of the earthquake source under the surface, the curvature of the Earth, the bending of rays due to velocity variations in the travel medium and the presence of elastic discontinuities. These variables have been taken into consideration

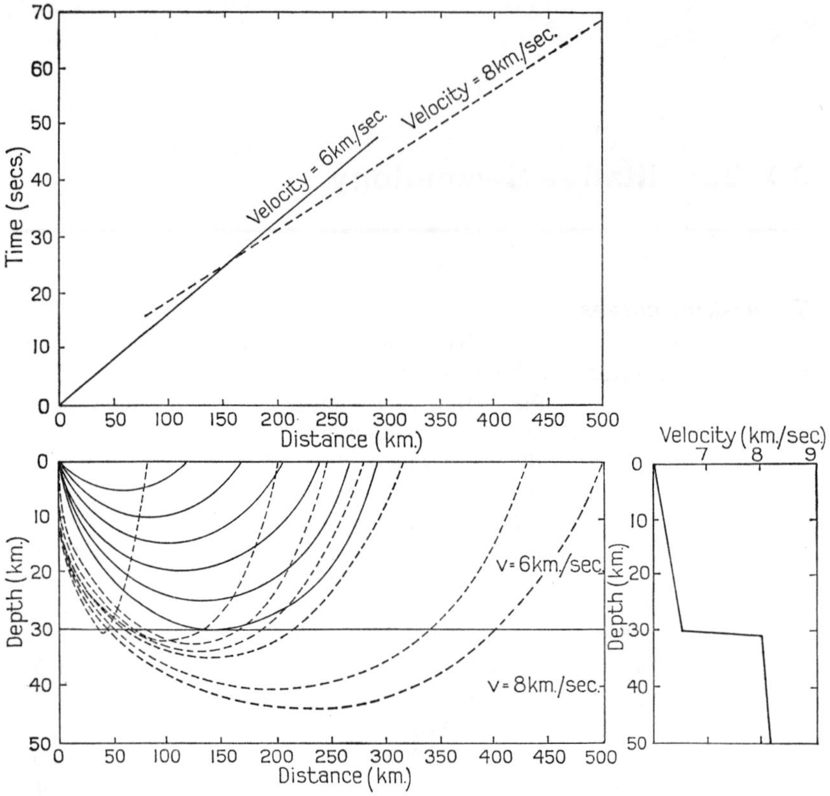

FIG 27 Principle of the travel-time curve. The seismic velocity structure of the two-layer case is shown at bottom right. At bottom left are the ray paths—the solid lines represent rays in the upper medium, and the dotted lines those controlled by the lower medium. The travel-time curve is shown above and is based on an actual computation. A vertical enlargement of 5 is used for the ray plot

and theoretical travel-time tables and curves, for various earthquake phases, epicentral distances and source depths have been computed, thanks to the monumental work of Jeffreys and Bullen on the one hand and Gutenberg on the other. Figure 28 shows a sample of Jeffreys-Bullen travel-time curves for a surface earthquake.

Naturally, such theoretical computations are necessarily based on simplified models of earth structure and assume homogeneous patterns all over the earth. Hence they are continually revised both by theoretical work to correct for factors like the ellipticity of the Earth and by data analysis for regional variations. As more and more actual travel-time data accumulate,

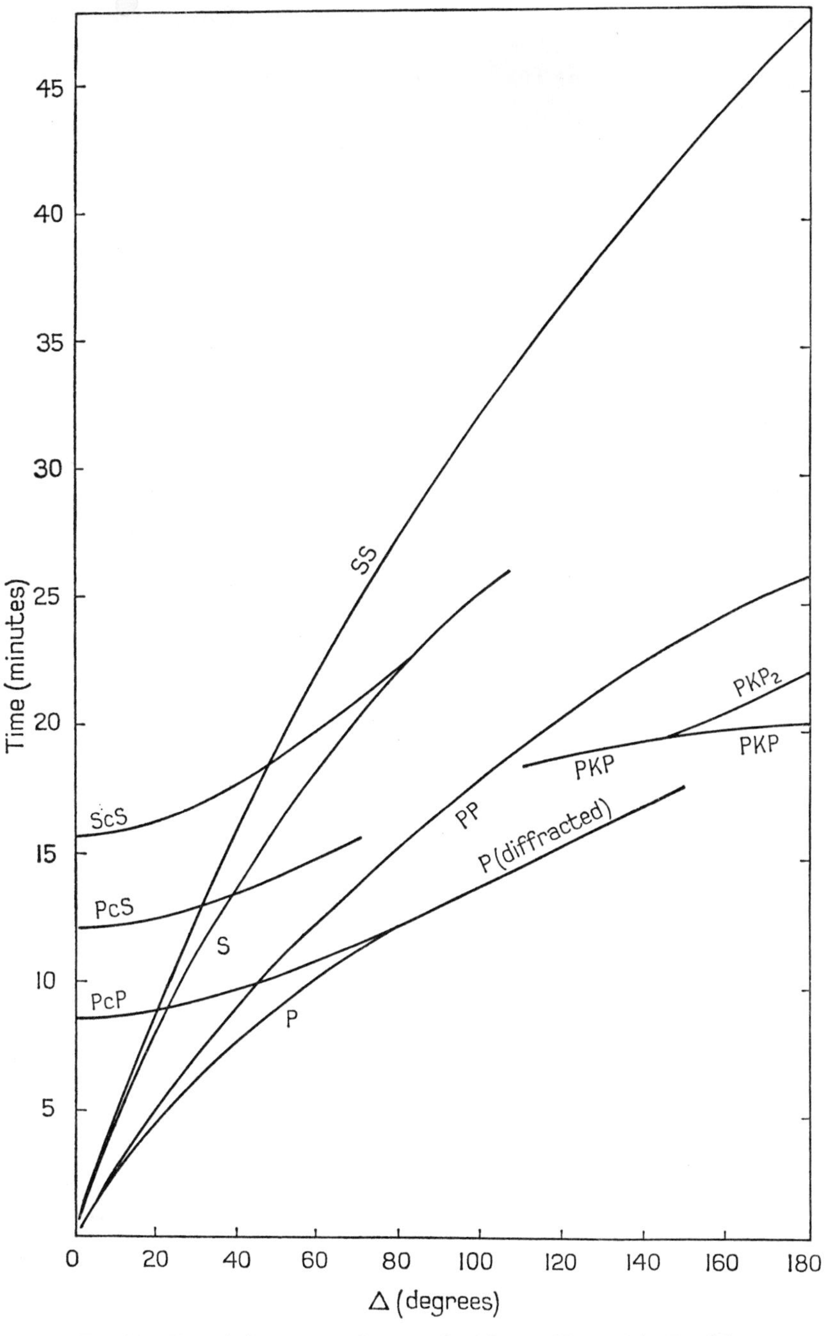

FIG 28 Travel-time curves for a surface focus. (From *The Earth* by Sir H. Jeffreys. Courtesy of Cambridge University Press.)

the tables are revised. The principle here is one of successive approximations. When an event is recorded by several stations, the parameters are computed using available tables. From the scatter in the values obtained a statistical fit is made to get best average values. Based on these values the travel-time tables are corrected. Thus theory and observations go hand in hand and the travel-time tables are continuously improved.

Epicentre and depth of focus

Epicentre

To begin with, the seismologist studies the seismogram, and carefully identifies all distinguishable phases. The next step is to determine the times of arrival of these phases, correct the measured times for clock error and plot the recorded times on the straight edge of a long strip of paper on the same time scale as the travel-time chart to be used. The strip is held vertical on the chart, parallel to the time scale and moved about until the plotted marks on its edge fit the travel times on the chart for the identified phases. The epicentral distance can now be directly read from the horizontal scale. The time of origin can be estimated by subtracting the travel time on the chart from the corresponding times plotted on the strip.

The technique, though it sounds simple, can be subject to many errors in the hands of an inexperienced person. The most important cause of error is mis-identification of phases. The only cure for this is experience. Local and regional variations in travel times also cause errors which can be identified and corrected as more and more data accumulate.

Several stations determining epicentral distances using the above method can pin-point the epicentre of the earthquake, by striking off arcs on a globe from the respective stations. A guiding principle in deciding which stations to use is consistency in computed origin times. In modern seismology, first arrival P-wave readings reported by several stations are used, together with the Jeffreys-Bullen or equivalent travel-time tables, to determine epicentre and depth of the earthquake. A trial and error method is used for the calculations, often in conjunction with fast digital computers.

The best epicentre locations depend on recorded times at several stations within a radius of about 300 km from the source. Usually in regions where there are a dense network of seismograph stations, the epicentre locations can be very accurate. It has been estimated that the U.S. Geological Survey's Seismic network for the study of earthquakes caused by California faults, can establish the location of events that occur within the network to an accuracy of about 1 km horizontally and 2 km vertically.

Depth of focus

Usually, most earthquakes occur above the Moho, a common average depth being 25 km. These are called *shallow earthquakes*. Travel-time information indicates only that the event is shallow, that is at depth between 0 and 50 km. However, it is known that earthquakes can occur at depths of as much as 700 km. Of the deeper ones, those down to depths of 300 km are referred to as *intermediate*, while those which occur at greater depths are called *deep*. Such earthquakes have some special phases which enable fairly accurate depth determinations to be made.

Deep focus earthquakes were discovered around 1922. Before that it was believed that all earthquakes were shallow. However, continued use of travel-time tables based on shallow earthquakes showed that, for a small percentage of events, travel times for certain phases like the *PKP* were abnormally smaller. The observations aroused considerable interest, and detailed study of such time discrepancies revealed the presence of deep focus earthquakes. The characteristics of deep-focus seismograms are now well known and corrected travel-time charts are available to allow for deep foci.

The following are some of the special characteristics of deep focus earthquakes.

(i) Small amplitude (or even absence) or surface waves.

(ii) Special *P*- and *S*-phases, caused by reflection at the earth's surface, close to the epicentre. These phases are referred to as *pP* and *sS* and are excellent depth indicators. In Figure 29, the mechanism of formation of these possible reflected *P*-wave phases from a deep-focus earthquake is illustrated. Corresponding *S* phases are also possible. The waves are assumed to have a constant velocity within the sphere.

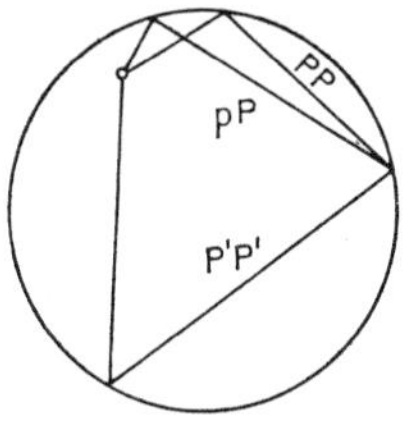

FIG 29 Deep-focus reflected phases (*P*-wave) in a sphere of constant velocity. (From *Elementary Seismology* by Charles F. Richter. W. H. Freeman & Co. Copyright © 1958)

(iii) For shallow earthquakes, effects like destructiveness for strong shocks and wave-amplitude for smaller shocks reduce sharply as one moves away from the epicentre. However, for deep focus earthquakes the reduction is at a smaller rate.

Magnitude of earthquakes

The strength of an earthquake is best estimated by measuring the ground motion it produces on the surface of the Earth. In the case of large earthquakes, of course, the strength can be determined by the visible effects they produce. Before instrumental siesmology became fully developed, such visible phenomena were the only way of judging how strong the earthquake was. Based on earthquake effects, several Intensity Scales were developed in the past. For many years, the most widely used was the scale set up by Rossi and Forel in 1878. A later scale was developed by Mercalli and a modified version of this is still in use. The Modified Mercalli Scale, as it is called, divides earthquakes into twelve progressively destructive categories. An abridged form of this scale is given in Table 2. The corresponding Rossi-Forel Scale numbers (approximate) are indicated in brackets.

TABLE 2

Modified Mercalli Scale for earthquakes. Corresponding Rossi-Forel scale numbers (approximate) are indicated in brackets. (From Bullen, *An Introduction to the Theory of Seismology,* Cambridge University Press)

I. Not felt except by a few under especially favourable circumstances. (R.F., I.)

II. Felt only by a few persons at rest, especially on upper floors of buildings. Delicately suspended objects may swing. (R.F., I–II.)

III. Felt quite noticeably indoors, especially on upper floors of buildings, but many people do not recognize it as an earthquake. Standing motor cars may rock slightly. Vibration like passing of truck. Duration estimated. (R.F., III.)

IV. During the day felt indoors by many, outdoors by few. At night some awakened. Dishes, windows, doors disturbed, walls make creaking sound. Sensation like heavy truck striking building. Standing motor cars rocked noticeably. (R.F., IV–V.)

V. Felt by nearly everyone, many awakened. Some dishes, windows, etc., broken; a few instances of cracked plaster; unstable objects overturned. Disturbance of trees, poles, and other tall objects sometimes noticed. Pendulum clocks may stop. (R.F., V–VI.)

VI. Felt by all; many frightened and run outdoors. Some heavy furniture moved; a few instances of fallen plaster or damaged chimneys. Damage slight. (R.F., VI–VII.)

VII. Everybody runs outdoors. Damage negligible in buildings of good design and construction; slight to moderate in well-built ordinary structures; considerable in poorly built or badly designed structures; some chimneys broken. Noticed by persons driving motor cars. (R.F., VIII−.)

VIII. Damage slight in specially designed structures; considerable in ordinary substantial buildings, with partial collapse; great in poorly built structures. Panel walls thrown out of frame structures. Fall of chimneys, factory stacks, columns, monuments, walls. Heavy furniture overturned. Sand and mud ejected in small amounts. Changes in well water. Disturbs persons driving motor cars. (R.F., VIII+ to IX−.)

IX. Damage considerable in specially designed structures; well-designed frame structures thrown out of plumb; great in substantial buildings, with partial collapse. Buildings shifted off foundations. Ground cracked conspicuously. Underground pipes broken. (R.F., IX+.)

X. Some well-built wooden structures destroyed; most masonry and frame structures destroyed with foundations; ground badly cracked. Rails bent. Landslides considerable from river banks and steep slopes Shifted sand and mud. Water splashed (slopped) over banks. (R.F., X−.)

XI. Few, if any, (masonry) structures remain standing. Bridges destroyed. Broad fissures in ground. Underground pipe lines completely out of service. Earth slumps and land slips in soft ground. Rails bent greatly. (R.F., X.)

XII. Damage total. Waves seen on ground surfaces. Lines of sight and level distorted. Objects thrown upward into the air. (R.F., X+.)

The intensity rating for a particular earthquake varies from place to place and depends on the distance and depth of focus, the nature of the ground and the substratum, the solidity of the buildings and the imagination of the reporting inhabitants. After an earthquake, equal intensity curves, called *isoseismal lines* (see later section) are drawn from the data collected.

Since the dawn of instrumental seismology, efforts have been made to employ instrumental recordings of seismic waves to compute the strength of the earthquake and to use the information to estimate the energy release at the source. In 1935 Richter set up a Magnitude Scale to enable comparison of California earthquakes with one another. Magnitudes were calculated on the basis of maximum amplitude of the waves as recorded by Wood-Anderson torsion seismographs. The standard was based upon the indications of such an instrument with a natural period of 0·8 sec, magnification of 2,800, and a damping of 0·8 critical. The *magnitude* M_L was defined as $M_L = \log_{10} A$, where A is the maximum amplitude in microns recorded by a standard seismometer at a distance of 100 km from the epicentre. Reduction of observed amplitudes at various distances to expected amplitudes at 100 km was made using empirical amplitude-distance tables. The scale was applied directly only to earthquakes of shallow depth of focus.

Richter applied the magnitude scale only to local earthquakes with epicentral distances less than 600 km. Later Gutenberg developed empirical tables to enable observations at distant stations, and on non-standard seismographs, to be reduced to magnitude values. The tables were also extended to cover earthquakes at various focal depths and to enable magnitude determinations using body wave and surface wave data independently.

Currently there are three types of magnitude determinations in use:

(i) Magnitude of local earthquakes (M_L) based on maximum wave-amplitudes recorded by standard Wood-Anderson seismographs.

(ii) Magnitude using teleseismic surface wave amplitudes (M).
The procedure to compute M is the same as that for M_L, except that the maximum amplitude (in microns) of 20 sec surface waves is used. Seismograph response corrections and travel distance corrections are applied to arrive at the correct value.

(iii) Magnitude using teleseismic body waves (m).
This estimate, developed by Gutenberg in 1945, enables use of observed amplitudes and periods of phases like *P*, *PP*, *S*, etc. From observation $\log_{10}(A/T)$ is computed, where A is the amplitude in microns and T the period in seconds. The values are corrected for distance and depth of focus. The advantage of this scale is that it can be used equally well for shallow as well as deep shocks, unlike M which is confined to shallow shocks, for deep focus earthquakes do not produce appreciable surface waves.

It became apparent, as data accumulated, that M_L and M could be treated as a continuation from local to teleseismic events. However, 'm' showed systematic deviation from the other two. The relation between 'm' and the other measures is continuously being revised as data becomes more precise and abundant. One relationship given by Richter in 1958 is as follows:

$$m = 2{\cdot}5 + 0{\cdot}63M$$

Energy of earthquakes

The main use of magnitude is to enable computation of energy release associated with earthquakes. Empirical relationships between magnitude and energy are numerous and frequently revised. A relationship currently in use is the following:

$$\log_{10} E = 11{\cdot}8 + 1{\cdot}5M$$

Here E is the energy release in ergs. Using this relationship the energy associated with the largest earthquake of the century, the magnitude 8·9 Columbia-Ecuador shock of January 31, 1906 is of the order of 10^{25} ergs, the same order as that estimated for an oceanic cyclone, and in a 100 megaton nuclear bomb.

3.5 Earthquakes—Their Causes and Distribution

Shallow earthquakes

Much of our knowledge of what happens in the vicinity of the focus of an earthquake has come from the work of H. F. Reid, who studied the effects of the Great San Francisco earthquake of 1906. He was the originator of the *elastic rebound* theory of earthquake mechanism, according to which the earthquake is not just a sudden happening, but is the effort of the earth to return to normal after it has been slowly strained for a long period of time. Thus the immediate cause of an earthquake is a sudden release of elastic strain energy accumulated across a pre-existing *fault* or fracture. The essence of the elastic rebound theory is that, strain energy builds up and, after a sufficient interval of time (measured in years and centuries), the blocks on either side of the fault are displaced relative to each other. Suppose parallel lines are drawn across a fault before the energy build up has taken place, as in Figure 30(*a*). As strain builds up, the lines get distorted (*b*), but the block surfaces still remain cemented together due to frictional forces. However, there comes a time, when the strain forces ex-

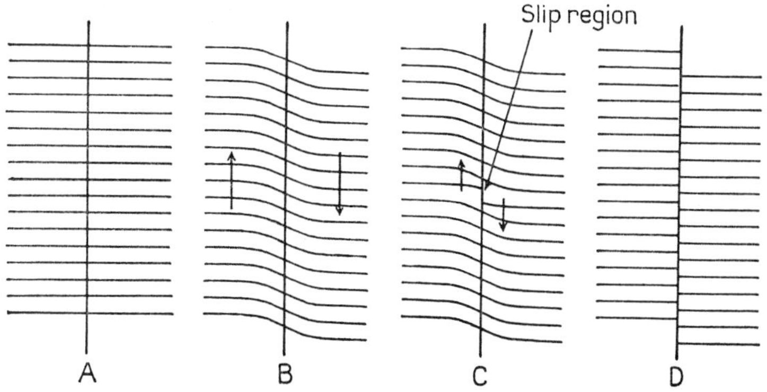

FIG 30 Illustration of the elastic rebound earthquake source theory. (Courtesy of H. Benioff and *Science* Vol. 143, pp. 1399–1406, 27 March 1964. Copyright 1964 by the American Association for the Advancement of Science)

ceed the restraints and there is a sudden slip or rebound (*c*). The slip at one point triggers off slips at adjacent points and this break is propagated along the fault (*d*). It is this sudden movement in opposite directions that causes the seismic waves. A simple experiment can be done to show the elastic rebound. A strip of tempered steel is tightly clamped at either end between wooden blocks. If the blocks are now moved parallel to each other but in opposite directions, the blade can be slowly bent to the breaking point. When the blade snaps, the broken ends whip back into straight pieces, but the ends are now considerably offset. The 'twang' corresponds to the earth tremor.

If the slip is in horizontal plane, the fault is designated as a *strike-slip*; if it is vertical, that is, if one block moves up and the other down, the fault is called *dip-slip*.

The closely watched, most studied fault on the earth's surface is the San Andreas fault, which extends along the whole length of California and for an unknown distance under the ocean to the north and south. This is a strike-slip fault and, according to some geologists, has been in existence for nearly twenty million years. The San Andreas fault must have fathered hundreds of earthquakes. Even during the past two hundred years it has moved violently half a dozen times. Of these movements two, the Tejon earthquake of 1857 and the San Francisco earthquake of 1906, were very violent.

The observed displacement across a fault varies from a few centimetres for very small earthquakes to several metres in large earthquakes.

The time taken for the elastic rebound to occur is not known accurately, but is estimated to very from a fraction of a second in smallest quakes to about ten seconds in the largest.

The horizontal extent of the break also varies with earthquake size. The largest observed was visible for over 330 km after the San Francisco earthquake, in the northern part of the San Andreas fault.

The rate of accumulation of elastic strain energy is known only for the San Andreas fault. Recent geodetic measurements conducted by the U.S. Geological Survey show that the fault blocks are moving at the rate of about 2–3 cms per year. This movement, though almost unperceptible, is really large in terms of a geological time scale. Considering the age of the fault, this could represent a total movement of the order of several hundred kilometres.

The depth extent of faults is estimated to be as much as the focal depth, that is, as much as 30–40 km.

The first clear evidence of faulting accompanying an earthquake was obtained in the Cutch earthquake of 1819. A most remarkable instance of a dip-slip fault was found after the Ancash earthquake of November, 1946 in the Peruvian Andes. Two parallel fault lines were found about three km apart, separating a zone which had subsided by some three metres. The largest unbroken trace of the fault extended for nearly five kilometres.

The elastic rebound theory enables computation of magnitude of the largest possible earthquake. The upper limit for magnitude is set by the fact that, once the strain equals the breaking strength of the rocks, faulting must take place. The upper limit of magnitude estimated using this idea works out to about 8·6. Thus any earthquake whose magnitude approaches this value is really great indeed. The Lisbon earthquake of 1755, the San Francisco earthquake of 1906, the Assam earthquake of 1950, the Chilean earthquake of 1960 and the Alaskan earthquake of 1963 are examples of such great earthquakes.

Deep focus earthquakes

The body wave pattern from deep focus earthquakes is similar to that of shallow earthquakes. Hence, though it looks as if they are also generated by the elastic rebound mechanism, there are certain difficulties. The chief problem is that rocks in the deep earth are under such high pressure that strain accumulation would shatter them before producing slip. Hence special agencies, like the presence of crushed or liquified matter at interfaces have been suggested to reduce the stresses required to cause slip and hence earthquakes. Another possibility is that deep earthquakes are caused by a sudden collapse of a small volume of rock at the focus, such as could be produced by a sudden change of state. The void would be filled by quick downward movement of overlying rock, causing an earthquake. Recent studies in Peru and California using strain seismographs tend more to support the second hypothesis.

Sources of strain energy

So far we have assumed that certain forces were at work causing the accumulation of elastic strain energy, which is later released as earthquakes. The question naturally arises, what are the sources of such energy?

A world map showing the position of the epicentres of all large earthquakes (of magnitude equal to or greater than 8), that occurred from 1904 to 1954, indicates that the earthquakes are not randomly distributed all over the Earth, but instead, are confined to two rather well defined regions: (1), the margins of the circum-Pacific continents, and (2), a band extending from Southern Asia to Portugal. In a substantial portion of the circum-Pacific arc, there is evidence to believe that the surrounding continents are rotating clockwise relative to oceanic rocks. If this is true, the strain accumulated by this rotation should account for the earthquakes. But what produces the rotation? The general concensus of opinion is that it is caused by giant convection currents within the mantle of the Earth.

The records also show that the earthquake belt passes through two large mountain regions, the Himalayan and Alpine. In the circum-Pacific belt also, mountain ranges form the general marginal pattern. The second possible source of earthquake energy is thus associated with the growth and decay of mountains, referred to as *orogenesis*. In geomechanics there is a principle of isostasy which suggests that different portions of the earth's crust are very delicately balanced; this accounts for the thin crust under the ocean, the deeper crust under continents and the roots under mountains. The situation is analogous to pieces of wood of different sizes and shapes floating on water. If all pieces are of the same density the bigger the block, the larger the part that sticks out above water and the deeper it goes underneath. The dynamics of earth formation demands such an isostatic balance, as otherwise a mountain without a root would exert gravitational attraction on surrounding matter and cause complex forces. Thus the material of the Earth is so arranged as not to rise or sink. However, the situation does not remain in such static equilibrium for ever. The weather slowly and steadily crumbles the mountain and rivers and glaciers continuously erode it. In terms of geological time, the mountains are getting lighter, upsetting nature's balance. Hence they tend to rise, creating tremendous forces that generate earthquakes.

A third proposed energy source comes from the theory of *continental drift*. The idea is that, originally, on the surface of the globe there was only one continent and one ocean. At a later stage the continent split up and the pieces began drifting away. What we see as several continents today is the instantaneous position of these floating, drifting fragments. The cause of earthquake generating forces lies in this tearing away and movement, according to the proponents of the continental drift theory. (See Section on Ocean Floor Spreading).

Lastly, mention should be made of the oldest hypothesis of earthquake generation. The idea is that the earth was once a molten hot body and, as it cooled, the outer surface formed into a crust. As the cooling continued, the top of the mantle also began to solidify, but in the process contracted also, producing compressional earthquake generating forces in the crust-mantle boundary. However, this theory is slowly being discarded as it seems more reasonable to assume that the Earth began as a cold body and is getting hotter, because of the radio-active heat generation of the rocks.

Aftershocks

After a large earthquake, there are generally many smaller shocks from almost the same origin. This is to be expected, since the disruption attending the major earthquake can cause secondary strain build up in surrounding rocks, bringing them at some point close to the stress at which fracture occurs. Also, a single earthquake need not release all the stored up strain energy at once. A classic example of an earthquake generating aftershocks is the Great Alaskan Earthquake of 1963. Several thousand aftershocks were recorded from this quake during the first month after the disaster.

Other types of earthquakes

It is generally believed that the majority of earthquakes are caused by elastic rebound. However, many other possible causes have been suggested, and there is evidence that some of them could have been the source of certain earthquakes. The following classification of earthquakes can be said to include most of the likely mechanisms:

(i) Tectonic

(a) Elastic rebound as described already.
(b) Sudden shearing during plastic flow as in deep focus earthquakes.

(ii) Volcanic

Earthquakes are often associated with volcanic eruptions. How volcanism can set off an earthquake is not clearly known, except in the case of volcanic explosions. The most famous recorded volcanic explosion was the eruption of Krakotoa, in the Sunda Strait between Java and Sumatra, in August 1883. The series of gigantic explosions from the volcano caused widespread destruction in the islands. The noise of the largest explosion was even heard in Australia!

Shallow shocks are frequently observed in the vicinity of volcanoes. These shocks are known to increase in frequency prior to an eruption and hence provide valuable data for predicting the catastrophe. For this purpose, volcano observatories are usually provided with low magnification seismographs to record the volcanic tremors.

(iii) Impact

The most important category of impacts giving rise to earth movements are the surface explosions caused by man, accidentally or intentionally. The explosions in Oppau, Germany, on September 21, 1921 was recorded throughout West Germany. In modern times nuclear bombs have energy release capabilities very close to medium earthquakes. The subject of nuclear bombs and what they have done to seismology is discussed in a later section.

Large meteorites striking the earth can produce appreciable earth-movements. The only documented instance of a meteorite crash capable of causing an earthquake is the great fall in Siberia on June 30, 1908. The huge air waves shook structures a hundred miles away from the impact area. They were recorded by British barographs. There is no report that the earth movement itself was felt, though the meteorite devastated a forest area of 15–20 km in diameter.

Geographical distribution of earthquakes

A sketch of the global distribution of large earthquakes has already been given in the discussion on sources of earthquake energy. This picture will now be developed in greater detail.

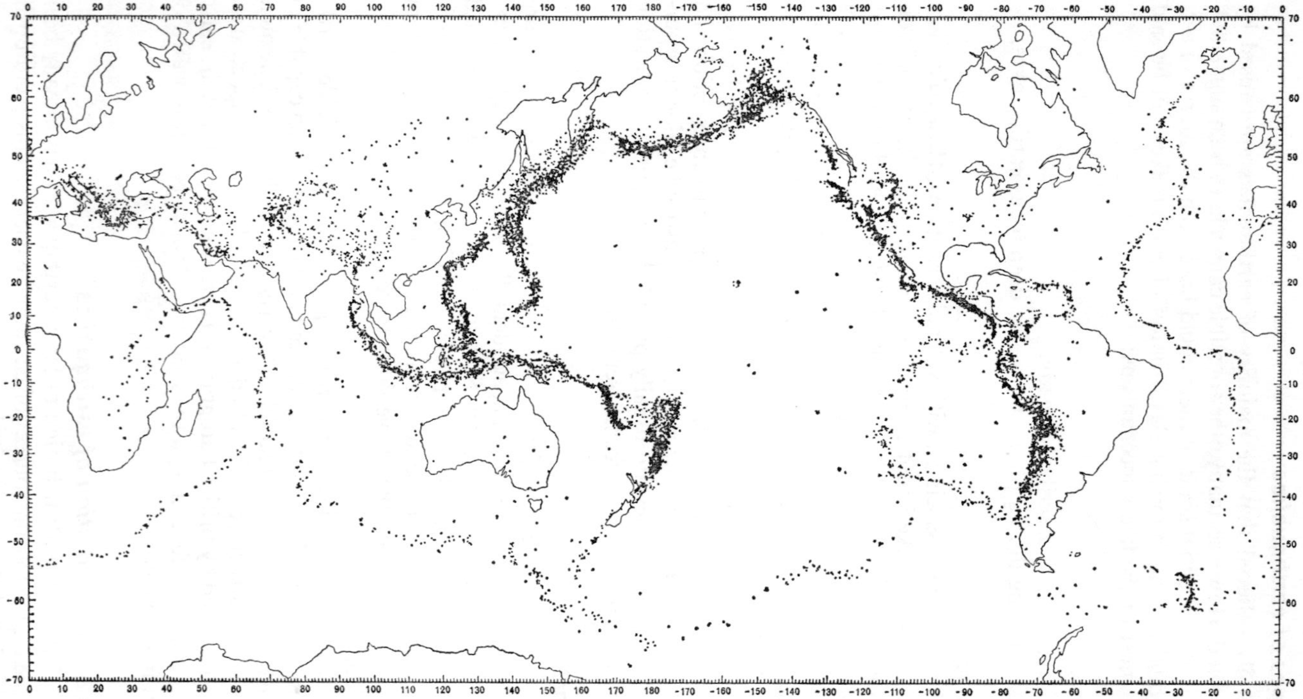

FIG 31 World seismicity map, 1961–67. (From Barazangi and Dorman, *Bull. Seism. Soc. America,* Vol. 59, No. 1, February 1969)

The geographical distribution of earthquakes over the world describes its *seismicity*. In fact early seismology was mainly concerned with making seismicity studies of particular areas by listing the shocks felt. Naturally the results were purely preliminary in view of the highly subjective data used. The credit for making the first proper seismicity map of the world goes to John Milne. From a humble beginning, of cataloguing Japanese earthquakes that had occurred between 295 B.C. and 1872 A.D., Milne prepared a detailed seismicity map of Japan using 8331 earthquakes actually observed during a period of eight years. The data used were very crude in the sense that they were based on massive information collected by voluntary observers. Later, when his world-wide seismograph network was set up, Milne prepared a world seismicity map.

Figure 31 shows a world-wide distribution of earthquakes (after Barazangi and Dorman, 1968). The data used cover the period 1961–1967, and were reported by the U.S. Coast and Geodetic Survey. Seismologically speaking, the most important subdivisions of the Earth's seismic belt are the following (after Richter):

(1) The Circum-Pacific belt
(2) The Alpide belt across Europe and Asia
(3) The Pamir-Baikal zone of Central Asia
(4) The Atlantic-Arctic belt
(5) The belt in the central Indian Ocean
(6) A zone in east Africa
(7) A wide triangular zone in east Asia between the Alpide and Pamir-Baikal zones
(8) Minor seismic areas, usually in regions of older mountain building
(9) The central basin of northern Pacific Ocean, almost non-seismic, except for the Hawaiian islands
(10) The stable central shields of continents, also nearly non-seismic.

The largest stable area of the Earth is considered to be the basin of the Pacific Ocean.

Stable continental areas, the so-called shields, exist in Canada, Brazil, the Baltic region, Africa, Central Asia, Arabia, India, and Australia.

The world seismicity map is by no means a conclusive record of seismic activity of the Earth. As more and more seismic data, from larger numbers of sensitive instruments, widely distributed over the globe, become available, smaller and smaller tremors are being detected. Thus even regions which were considered to be completely non-seismic are found to be otherwise. Hence the definition of seismicity needs to be modified to take into account even the smallest earthquakes, and statistical compilations have to be revised accordingly. It may well be that technically speaking there is no non-seismic area anywhere on the surface of the globe!

An interesting feature of the seismicity of the Earth is the fact that shallow and deep focus activities are not exactly superimposed and there is a pattern in their occurrence. In general earthquakes deeper than 100 km seem to occur in the vicinity of deep-oceanic trenches and island arcs, (the Hindukush, Assam and Mediterranean earthquakes are exceptions). Extensive compilations of deep-focus earthquakes in the New Zealand, Tonga-Kermadec, Japan and Kamchatka areas (Gutenberg and Richter, 1954) and recent studies by Sykes (1966) have shown that the spatial distribution of intermediate and deep earthquakes follow a well-defined pattern. For example in the Tonga-Kermadec region, where the trench is nearly 30 000 feet deep with a chain of islands running parallel to the trench to the west, the earthquakes are shallow between the trench and the island chain. The earthquakes tend to get deeper and deeper as their epicenters get farther and farther to the west of the island arc till a maximum depth of about 600 km is reached. The foci distribution seems to suggest a rough plane, dipping down towards the continental side of the arc. This thin strip containing deep earthquakes, sometimes referred to as the 'Benioff Zone', occurring in regions of severe tectonic disturbance, has puzzled geophysicists for a long time. Recently, however, the concept of ocean-floor spreading seems to be able to account for the deep-focus earthquake activity, at least in most of the regions (See later section on 'Continental drift, ocean floor spreading and the new global tectonics').

In summarizing, we can say that the major earthquake zones of the world seem somehow to be associated with island arcs, trenches and mountain ranges and that the cause of their existence probably is also the cause of earthquakes.

Some great earthquakes

It is estimated that the Earth is probably shaken by something of the order of one to ten million earthquakes every year. However, our seismographs are not sufficiently sensitive nor widely distributed enough to record every one of them. With the rapid advancement in seismic instrumentation that was made in the nineteen sixties, it can be hoped that we should soon be able to hit the million mark.

The exact number of earthquakes that occur is a fact of academic interest only. To a person who has actually been in a major earthquake, the term means something more than a wiggly line recorded by an instrument. He sees it in terms of destruction and suffering. To him the academic questions of seismology mean nothing, and every earthquake is an awesome 'act of God'. Thus really to understand the inter-relation between seismology and our daily life on earth, one has to study the effects of the notorious earthquakes of the past. We proceed to describe some of the historical earthquakes, selecting the events with a view to showing different types of earthquake effects. In understanding them it should be remembered that magnitude and destruction are not always synonymous, as the latter depends largely on where the earthquake occurred.

Lisbon earthquake, 1755

The Lisbon earthquake of 1755 can be considered as a spectacular event of the near past. There were three principal shocks at approximately 9:40, 10:10 and noon on the morning of November 1. Shocks were felt over an extensive area covering south western Europe and north eastern Africa. Lisbon suffered severely, about 60 000 out of its 235 000 inhabitants being killed. A noteworthy feature was that the part of the city which was built on soft sediments was almost completely destroyed, while the part on rock foundation suffered less serious damage. When the earthquake struck, rapid small vibrations were felt first. About thirty seconds later, violent rapid movements came and lasted for about two minutes. A minute later, violent upward movements followed and lasted about two and a half minutes. It is presumed that these were the *P*-, *S*- and surface-wave phases respectively. The earthquake set up *seiches* in ponds and lakes in Portugal, France, Italy, Holland, Switzerland, Norway and Sweden. (*Seiches* are sloshing movements in enclosed bodies of water. Such bodies have their own natural frequency of oscillation and, when excited by earthquake waves of the same frequency, they resonate, producing large amplitude seiches.) Historically, the Lisbon earthquake is very important as it produced the impetus for a scientific study of earthquakes.

New Madrid earthquakes

These were a series of earthquakes that shook the small town of New Madrid in U.S.A. again and again for about a year, and produced drastic changes on the surface of the earth. Their severity is estimated, not in terms of lives lost, but in terms of the large-scale effects in the area itself and the Mississippi river. The first quake struck at 2:00 a.m. on the morning of December 16, 1811. The inhabitants, about 800 in number, were woken up by the creaking of their log houses, by the furniture being thrown around and by crashing chimneys. Morning brought more shocks, to be followed by more day after day. When the series had ended, their effects were noticeable over an area of about 5,000 square miles. The most spectacular effects were confined to the Mississippi river and its tributaries. Some parts of the river bottom subsided to form permanent swamps and lakes. Other parts were elevated. The pressures built up by the movements of the river bed were released in the form of fountains of air, water and sand gushing out to heights of about fifteen feet. Large landslides occurred by the river banks, and several islands in the Mississippi river vanished completely. Temporarily the river changed its course. Large waves moving along the surface waters did great damage. The town was almost completely destroyed by the collapse of the river banks and the waves.

Indian earthquakes

The earthquake of June 12, 1897 which occurred in Assam, India, is also referred to as the Great Indian earthquake or the Assam Earthquake. This earthquake is important, not only because it is estimated to have had a magnitude of 8·7, but also because it was the first scientifically well documented great earthquake. The credit for this goes to R. D. Oldham, who was the Director of the Geological Survey of India at the time of the earthquake. He carried out a thorough investigation of the earthquake effects and his monograph is a very valuable source book in seismology. Oldhams' account of the earthquake appears in Elementary Seismology (Richter). About 1,500 lives were lost in the quake and the city of Shillong completely destroyed. It has subsequently been rebuilt and is an active centre for seismological studies in India.

The India-Pakistan subcontinent has been subjected to the following severe earthquakes during the past 150 years, besides the Assam earthquake:

(1) Cutch earthquake, June 16, 1819.
(2) Kangra earthquake, April 4, 1905. This has been assigned a magnitude of about 8·6.
(3) Quetta earthquake, May 30, 1935. Though of magnitude of only 7·6, this quake laid waste the city of Quetta and killed about 30 000 people.
(4) Assam and Tibet earthquake of August 15, 1950. The shock of magnitude around 8·7 was more damaging in terms of property loss than the 1897 earthquake.

San Francisco earthquake

At 5.12 a.m. on April 18, 1906 the city of San Francisco experienced a sharp tremor, followed by a jerky roar of collapsing man-made structures and a dull booming from the earth itself. Within a minute, the shaking had ceased, but the earthquake earned world-wide notoriety because of the fire that followed, and which did most of the damage. The losses caused by the fire were estimated to be of the order of 400 million dollars. The earthquake itself caused damage worth less than five per cent of this. The reported loss of life was about 700.

However, the importance of the earthquake arises from the fact that it left visible clues as to its cause, and a thorough investigation followed. Visible displacement of the San Andreas Fault was a spectacular feature of this earthquake. The fault broke over a length of about 200 miles, and every town and village within twenty miles on either side of it was damaged. Fences and roads crossing and fault were off-set; water pipes were disrupted. The displacement of the fault was purely horizontal, the Pacific side moving north relative to the continental side. The displacement reached a maximum of 21 feet near Tomales Bay and died to almost nothing at both ends.

In sixty seconds the earthquake crumbled to pieces the pride of the city, the six million dollar city hall. Overturned stoves and blocked chimneys were the causes of the fire. Fire fighting was made difficult as the water mains broke and no water was available. Pumping water from every available source, the city fought the fire for three days before bringing it under control. During this period the fire consumed thirty schools, eighty churches and convents, the homes of a quarter million people, and 450 lives.

Kwanto or Tokyo earthquake

On September 1, 1923, an earthquake of magnitude 8·2 devastated the cities of Tokyo and Yokohama. The focus of the earthquake was under Sagami Bay. The earthquake manifested itself with a roar followed by frantic shaking. As in San Francisco, fire started and fire fighting equipments were destroyed. When it was over, Yokohama had lost 27 000 lives, 40 000 people were injured and 70 000 homes destroyed. In Tokyo, 100 000 people were dead, 40 000 injured and 400 000 houses lost. During the month of September the Tokyo seismograph station recorded 1256 shocks of which 237 were felt. The floor level of Sagami Bay changed suddenly, releasing a giant *tsunami* or seismic sea-wave (see earthquake effects) which piled about thrity feet of water along the shores and caused much destruction. Some visible faulting occurred to the north and east of Sagami Bay.

Grand Banks earthquake

This earthquake, which occurred on November 18, 1929 under the ocean, on the continental slope, south of Newfoundland and east of Nova Scotia, did not cause much destruction. It is interesting because it provided direct evidence on the existence of *turbidity currents*, which are fast flowing currents composed of high density mixture of sand, pebbles and sea water, moving down continental slopes. The indication of the current was provided by the breaking of submarine cables in the vicinity of the epicentre in a total of 28 places. The cable breaks did not all occur at the same time, but one after another corresponding to increase in distance, as if some fast moving agency was snapping them. From the times of cable breaks is was estimated that the current was moving at a speed of around 50 knots. Later investigations showed that, on the steep part of the continental slope, the cables were unburied whereas on the flat ocean floor they were buried under deposits of 'sharp and small pebbles'. The hypothesis advanced for the origin of turbidity currents is that, in the vicinity of the earthquake focus, slumping and sliding of material occurred which mixed with sea water and moved as a powerful current.

Two ships which were in the epicentral area felt the quake as violent vibrations, as if the ships had hit submerged material. The earthquake also caused a *tsunami* which was recorded by tide guages as far off as the Azores and Bermuda.

Recent earthquakes

Since 1960 several destructive earthquakes have occurred and the havoc caused by them is still raw in our memory. We now proceed to discuss three of the major ones.

Agadir earthquake

This earthquake, which struck the city of Agadir, Morocco, just before midnight on February 29, 1960, almost completely destroyed the city and wiped out about a third of its population of 33 000. The earthquake was only 1/6300 as strong as the San Francisco earthquake, but the epicentre was only about two miles outside the city limits. The chief lesson that was learned, in a very painful way, from the earthquake was that the buildings should have had earthquake resistant construction. It was unfortunate that a past disaster in 1751 which had wrecked the city almost entirely, had been completely forgotten.

Great Chilean earthquake

1960 was a terrible year for Chile as it experienced a series of earthquakes. The series began with one large shock followed by a series of aftershocks on May 21, 1960. Next day at 3.00 p.m. a large earthquake, twenty times stronger than the initial shock, struck. The earth moved in smooth undulating motions, of period 10–20 seconds, for more than three minutes. The shocks produced such serious damage that it will take years before the economy of the country recovers completely.

The Chilian earthquake produced a spectacular tsunami. Shortly after the main earthquake, the inhabitants residing in coastal areas noticed that the sea had begun to recede rapidly. The authorities knew that this was the first ominous sign of the dreaded 'tidal wave' and evacuated the people to higher grounds. In about ten to twenty minutes, the sea returned, thundering into the interior with twenty foot waves, washing away and crushing everything in its path. The sea receded and advanced several times during the afternoon, the third and fourth waves being the highest. Many people lost their lives as they returned to watch after the sea receded for the second time, thinking that the worst was over. The 'tidal wave' travelled across the globe at speeds of the order of 400 miles per hour. At Hilo in the Hawaiian group, 35 foot waves were recorded. Twenty-two hours after the earthquake, the tsunami reached the Japanese coast, more than 10 600 miles away and caused about 70 million dollars worth of damage.

Another spectacular effect of the earthquake, though not visible to the eye, was that the Earth as a whole was set into oscillations which lasted for about a month. Being an elastic body, it was known that the Earth could go into normal mode oscillations (see later section), but the 'ringing' excited by the Chilean earthquake lasted for about a month and was recorded by very sensitive seismographs. The data provided very valuable material for the study of the structure of the Earth.

The earthquake produced extensive damage in the form of landslides, cracks on the ground, seiches, ground subsidence and rising, flooding, etc. One volcano (the Ruyehue) which has been dormant since 1905, erupted.

Prince William Sound Earthquake, Alaska

On March 28, 1964, an earthquake of magnitude 8·4 occurred under the Prince William Sound (bay), about 75 miles from Anchorage city in Alaska. The loss of life was estimated to be about 115. The earthquake caused property damage of the order of $500 000 000. Spectacular landslides and a giant tsunami were produced by the earthquake.

Summary of earthquake effects

The description of some of the historical earthquakes given in the previous pages, provides some picture of the possible effects of large earthquakes. The Mercalli scale of Intensity, given earlier in Table 2 provides a scientific classification of some of the most obvious effects, useful in determining the strength of an earthquake. The intensity estimates associated with an earthquake are usually plotted as an isoseismal map, which consists of contours of equal intensity over the earthquake area. Even though individual intensity values may be over-estimated the average isoseismal map gives a fairly realistic picture of earthquake effects. The irregularities on an isoseismal map tend to point out areas of poor soil or of poor building construction. Figures 32(a) and (b) show the isoseismal maps of the San Francisco earthquake. The dotted line indicates the San Andreas fault.

The *primary* effects of earthquakes are due to faulting and the *secondary* effects due to shaking caused by the passage of elastic waves. Table 3 gives a summary of the main earthquake effects.

Tsunami, also called seismic sea wave, is Japanese for a large wave entering a harbour. It is not clearly known how these waves are caused by an earthquake. When the tsunami originates in the vicinity of the earthquake

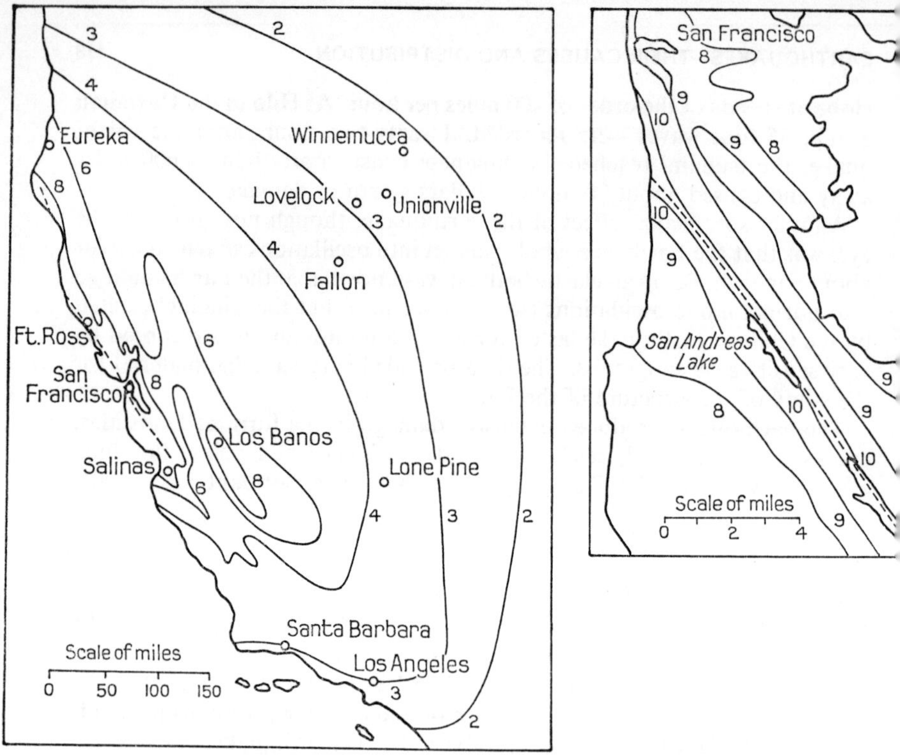

FIG 32 (a) and (b) Isoseismal maps of the San Francisco earthquake. The numbers denote intensity in modified Mercalli Scale. The dotted line represents the San Andreas fault. (From *Introduction to Geophysics* by Howell. Used by permission of McGraw-Hill Book Company)

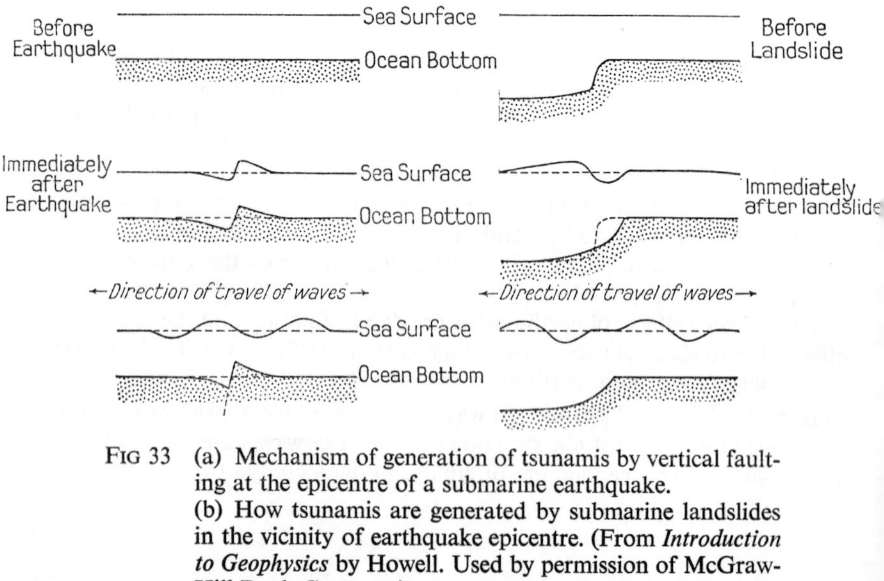

FIG 33 (a) Mechanism of generation of tsunamis by vertical faulting at the epicentre of a submarine earthquake.
(b) How tsunamis are generated by submarine landslides in the vicinity of earthquake epicentre. (From *Introduction to Geophysics* by Howell. Used by permission of McGraw-Hill Book Company)

TABLE 3

Summary of earthquake effects. (From *Elementary Seismology* by Charles F. Richter, W. H. Freeman & Co. Copyright © 1958)

Effects on	Primary	Secondary Permanent	Secondary Transient
Terrain	Regional warping, etc. Scarps Offsets Fissures, mole tracks, other trace phenomena Elevation or depression of coasts, changes in coast line	Landslides (slumps, flows, avalanches, lurches)[1, 2, 3, 4,] Secondary fissures[3] Sand craters[5] Raising of posts and piles	Visible waves[2] Perceptible shaking
Water (ground and surface)	Damming; waterfalls; diversion Sag ponds Changes in wells, springs		Changes in well levels Earthquake fountains[5] Water over stream banks Seiches Tsunamis Seaquakes
Works of construction	Offsets, and destruction or damage by rending or crushing; buildings, bridges, pipelines, railways, fences, roads, ditches	Most ordinary damage to buildings, chimneys, windows, plaster	Creaking of frame Swaying of bridges and tall structures
Loose objects		Displacement (including apparent rotation) Overturning, fall, projection (horizontal or vertical)	Rocking Swinging Shaking Rattling
Miscellaneous		Clocks stop, change rate, etc. Glaciers affected Fishes killed Cable breaks	Nausea Fright, panic Sleepers wakened Animals disturbed Birds disturbed Trees shaken Bells rung Automobiles, standing or in motion, disturbed Audible sound Flashes of light?

[1] Earth flows properly belong with water phenomena.
[2] Landslides may produce damage to works of construction.
[3] Landslides and secondary fissures may produce the effects on terrain and on surface water listed as primary.
[4] Classification of landslides from California Earthquake Commission, Vol. 1, pt. 2, p. 385.
[5] Production of sand craters and fountains is a single phenomenon.

epicentre, it is possible that it is generated by vertical faulting on the ocean floor suddenly sucking in the water surface and producing a seismic sea wave. Another possible cause is a submarine landslide, caused by the earthquake, but not necessarily in the vicinity of the epicentre. These two tsunami causing mechanisms are illustrated schematically in Figures 33(a) and (b) respectively.

The period of oscillation of the tsunami is of the order of an hour, the wavelength several hundred miles, speed $\sqrt{gh}$, where g = acceleration due to gravity, and h = ocean depth in the region over which the wave is passing. Over deep ocean the wave can travel at speeds of the order of 600 miles per hour. When the giant wave breaks over the coast line, the water piles up causing destruction.

Illusions and realities

Sea quakes often kill fishes in large numbers, the shock waves being the cause. There is a legend in seismology that animals can *sense* impending earthquakes and become uneasy. The only possible realistic explanation to this is that the animals feel the small foreshocks, not noticed by persons. During an earthquake animals show their disquietude strongly. Horses snort, cattle stampede, dogs bark and whine and cats spit. After the Long Beach earthquake it was reported that several cats did not return to their homes for several days.

Human reactions to earthquakes are diverse. Some are reported to become sick, some are 'thrown out of bed,'—probably as a nervous reaction rather than by the shock wave. There are accounts of people seeing waves on the ground. It is possible that such visions are purely illusions. On the other hand, the possibility of standing waves, as on strings in a familiar laboratory experiment, on the surface of the earth cannot be entirely ruled out. Some report to have seen flashing or presistent lights during earthquakes. These could be real arcs across broken electric wires, thunderstorms, displaced blocks discharging static electricity but, most probably, illusions.

The sounds associated with an earthquake are due to crashing structures in built up areas. However, in open country it is reported that low frequency sounds comparable to thunder, gunfire or heavy traffic can be heard. The sounds are probably caused by transference of energy from ground P-waves to air shockwaves.

The hot and sultry earthquakes weather is a fable. However, there are accounts that natives of Assam can sense earthquake weather and come out of their houses.

How does it feel to be in an earthquake? The author, fortunately, has never been in a major one. Reports show that the general effect in a building is as if a truck had backed into it, followed by shaking and rattling. Buildings appear to sway back and forth. If the shaking is violent, it is impossible to walk or even stand. In cars the sensation is like driving on a bumpy road, steering becomes difficult.

Contrary to popular belief, it is not always safe to rush out of doors during an earthquake, as the streets may be dangerous with falling masonry and flying debris. The best thing to do is to stay calm and, as soon as it is safe to move about, to put out all open flames.

Some earthquake statistics

By studying world seismicity over a long period of time it has been statistically estimated that the number (N) of earthquakes in any magnitude level is roughly ten times that at one level higher. This relationship has been expressed mathematically by Richter in the following formulae:

$$\log_{10} N = 7{\cdot}81 - 0{\cdot}58M, \qquad \text{for } M < 7{\cdot}3$$
$$\log_{10} N = 9{\cdot}1 - 1{\cdot}1M, \qquad \text{for } M > 7{\cdot}3$$

These formulas are empirically established with data from one particular region. Naturally, it is very difficult to arrive at any sensible relationship on a global basis. Extrapolating from the above relations it can be seen that about one to ten million shocks may be occurring every year.

Table 4 gives the number of large earthquakes that occur in various magnitude ranges. The data used covers a period of 38 years, 1918–1955.

TABLE 4

Number of large earthquakes that occur in various magnitude ranges. (From *Elementary Seismology* by Charles F. Richter. W. H. Freeman & Co. Copyright © 1958)

Magnitude Range	Shallow (0–70 km)	Intermediate (70–300 km)	Deep (300–600 km)
8·6 and over	9	1	0
7·9–8·5	66	8	4
7·0–7·8	570	214	66

The frequency of intermediate and deep shocks generally decreases with increasing depth. No shocks are known to have occurred at depths below 700 km.

Table 5 is a list of the large earthquakes of the world from the fifteenth century onwards. Available details on these earthquakes are also given.

TABLE 5

List of large earthquakes from 15th century onwards (Tazieff, *When the Earth Trembles*, Harcourt, Brace & World Inc., and Rupert Hart-Davis Ltd)

Date	Place	Death roll	Magnitude
1456	Naples	30 000	
1556	Shensi	830 000	
1716	Algiers	20 000	
1755	Lisbon	60 000	
1759	Baalbek	20 000	
1783–86	Calabria	60 000	
1819	Kutch (INDIA)	1,800	
1883 July 26	Ischia (ITALY)	2,300	
1891 Oct. 28	Mino-Owari (JAPAN)	7,300	
1897 June 12	Assam	1,542	8·7
Aug. 5	38°N 143°E		8·7
Sept. 20	6°N 122°E		8·6
Sept. 21	6°N 122°E		8·7
1899 Oct. 10	Yakutat (ALASKA)		8·6
1902 Aug. 22	40°N 77°E		8·6
1905 Apr. 4	Kangra (INDIA)	19 000	8·6
July 23	49°N 98°E		8·7
1906 Jan. 31	Colombia		8·9
Mar. 17	Formosa	1,250	
Apr. 18	San Francisco	700	8·3
Aug. 17	Valparaiso		8·6
1908 Dec. 28	Messina	82 000	7·5
1911 Jan. 3	Tien Shan		8·7
1915 Jan. 13	Avezzano (ITALY)	30 000	
1917 May 1	29°S 177°W		8·6
June 26	15°S 173°W		8·7
1920 Oct. 16	Kansu	180 000	8·6
1923 Sept. 1	Kwanto (JAPAN)	140 000	8·2
1927 Mar. 7	Tango (JAPAN)	3,000	7·9
1929 Mar. 7	51°N 170°W		8·7
1930 Nov. 26	Ind (JAPAN)		
1933 Mar. 3	Sanriku	3,000	8·9
1935 May 31	Quetta (BALUCHISTAN)	30 000	
1938 Feb. 1	5°S 130°E		8·6
1939 Jan. 25	Concepción (CHILE)	25 000	8·3
Dec. 26	Turkey	25 000	7·9
1940 Nov. 10	Bucharest	1,000	
1942 Aug. 24	15°S 76°W		8·7
1943 Sept. 10	Tottori (JAPAN)	1,400	7·4
1944 Jan. 15	San Juan (ARGENTINA)	5,000	
1946 Nov. 10	Ancash (PERU)	1,500	7·4
1948 June 28	Fukui (JAPAN)	5,300	7·3
Oct. 4	Turkmenistan (PERSIA)	3,000	7·3

TABLE 5—*cotinued*

List of large earthquakes from 15th century onwards (Tazieff, *When the Earth Trembles,* Harcourt, Bruce & World Inc., and Rupert Hart-Davis Ltd)

Date	Place	Death roll	Magnitude
1949 Aug. 5	Ambato (ECUADOR)	6,000	6·75
1950 Aug. 15	Assam	1,526	8·7
1951 May 6	San Salvador	4,000	
1952 Mar. 4	Tokaichi (JAPAN)	600	8·6
1953 Aug. 9–13	Ionian Islands	500	7 (a long, violent crisis)
1954 Sept. 9	Orléansville	1,250	6·7
Mar. 29	Granada (SPAIN)		7 (depth 400 miles)
1955 Feb. 17	Lipari (ITALY)		(depth 294 miles)
1956 July 9	Santorin (GREECE)	53	7·7
1957 Dec. 4	Altai (OUTER MONGOLIA)	20	8·6
Dec. 13	Kurdistan	2,000	7·25
Apr. 25	Thessaly (GREECE)		7·25
1960 Feb. 29	Agadir (MOROCCO)	10 000	
Apr. 24	Lar (PERSIA)	1,000	5·75
May 22	Chile (CHIEF SHOCK)	5,000–10 000	8·9
1963 July 26	Skopje (YUGOSLAVIA)	2,000	5·4
1964 Mar. 28	Alaska (U.S.A.)	115	8·4
1965 May 3	El Salvador	110	6·3
1966 Aug. 19	Turkey	2,394	6·7
Oct. 17	Near Coast of Peru	110	7·8
1967 July 22	Turkey	173	7·3
Dec. 10	Koyna (INDIA)	177	7·5
1968 Jan. 15	Sicily	252	5·4
Aug. 1	Luzon (PHILIPPINE ISLANDS)	270	7·3
Aug. 31	Iran	11 000	6·0
Sept. 1	Iran	2,000	5·9

Although the number of earthquakes increases rapidly with decreasing magnitude, the bulk of the strain energy is released in the few large shocks. Dr. Benioff has made a study of the strain energy release on a global scale, during this century, using the large shallow shocks. Figure 34 shows Benioff's curve of total global strain release as a function of time. Note the step-like formation of the curve, bursts of energy release followed by flat

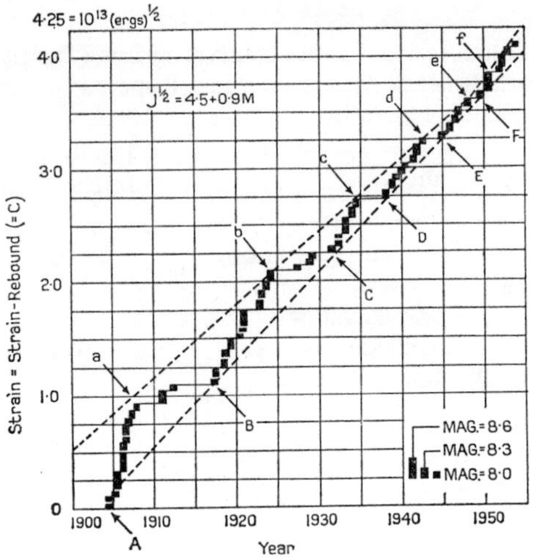

FIG 34 Global strain release with time. (From Geological Society of America, *Special Paper No.* 62). The ordinate is related to the square root of the total energy release.

regions indicating quiet periods. The bursts of activity are shown in figure by the following portions of the curve:

Aa	1904–1907
Bb	1917–1924
Cc	1931–1935
Dd	1938–1942
Ee	1945–1948
Ff	1950

The interesting feature of this important diagram is the successive decrease in the intensity of the bursts since the remarkably active 1900s. It is as though the earth was slowly settling down after a big disturbance, until 1948. Do the large earthquakes of the 1960s show the beginning of a second cycle and, if so, how many more disastrous earthquakes can we expect? This question, probably, has no immediate answer.

International study of earthquakes

Seismology has mainly depended for its growth on international co-operation. Many of the principal advances in the science have been made by the world-wide collection and analysis of data from a single large earthquake. Fortunately, the international courtesy required for exchange of data and ideas to facilitate such studies has existed in spite of wars and political barriers.

The first person to realize the need for international collection of seismic data was John Milne. Under his persuasion, in 1898, the British Association for the Advancement of Science recommended that a uniform network of seismographs be set up throughout the world. As a start, observatories in the British Empire were provided with Milne seismographs. Several national and regional networks followed, thanks to private groups and dedicated scientists. The progress of world-wide instrumentation has depended to a large extent on the development of seismographs themselves. Unfortunately, the non-standard nature of the instrumentation resulted in diverse types of seismographs producing different types of records on smoked paper and photographic sheets moving at various speeds. Despite these limitations, a great deal of knowledge about the earth's structure and earthquakes is derived from the non-uniform data.

Each seismic observatory usually has a seismologist, who reads the seismograms and issues bulletins giving details such as principal phases, their arrival times, amplitudes and periods of the waves, for the events that are recorded. The bulletins are distributed to all interested users. Sometimes national or regional observatories join together and make epicentral and magnitude determinations as well.

The U.S. Coast and Geodetic Survey acts as a co-ordinating agency for world-wide observations. It receives, telegraphically, readings from key stations all over the world for every recorded event. The data are collected in Washington, D.C., and a preliminary determination of epicentre is made. More recently, a digital computer has been used to digest the data and print out epicentre and other relevant information in a standard format. The result is the well-known Preliminary Determination of Epicentre Card, containing information such as origin time of event, latitude and longitude of epicentre, depth of focus and magnitude. Such cards are mailed to all bona fide users on the mailing list of the Survey. The survey also publishes detailed bulletins for later circulation.

Because of the destructiveness caused by tsunamis, the U.S. Coast and Geodetic Survey also operates a Tsunami Warning Service for the Hawaiian group of islands and the Pacific coast. The headquarters of the service is in Honolulu. A seismograph watch is maintained for 24 hours of the day for possible submarine earthquakes. Any suspicious event is cross-checked with other seismological observatories, tide-gauge stations, etc., using rapid communication facilities provided by governmental agencies. If a tsunami is suspected, sufficiently advance warnings are issued and necessary steps are taken to protect lives and property. It is hoped that the Service will be duplicated in other parts of the world as well.

Two other organizations, which operate much more slowly, are supported by the International Union of Geodesy and Geophysics. The first of these is the Bureau Central International Seismologique in Strasbourg, France. This organization collects data from stations all over the world, including stations that are slower in supplying data. The observations, put

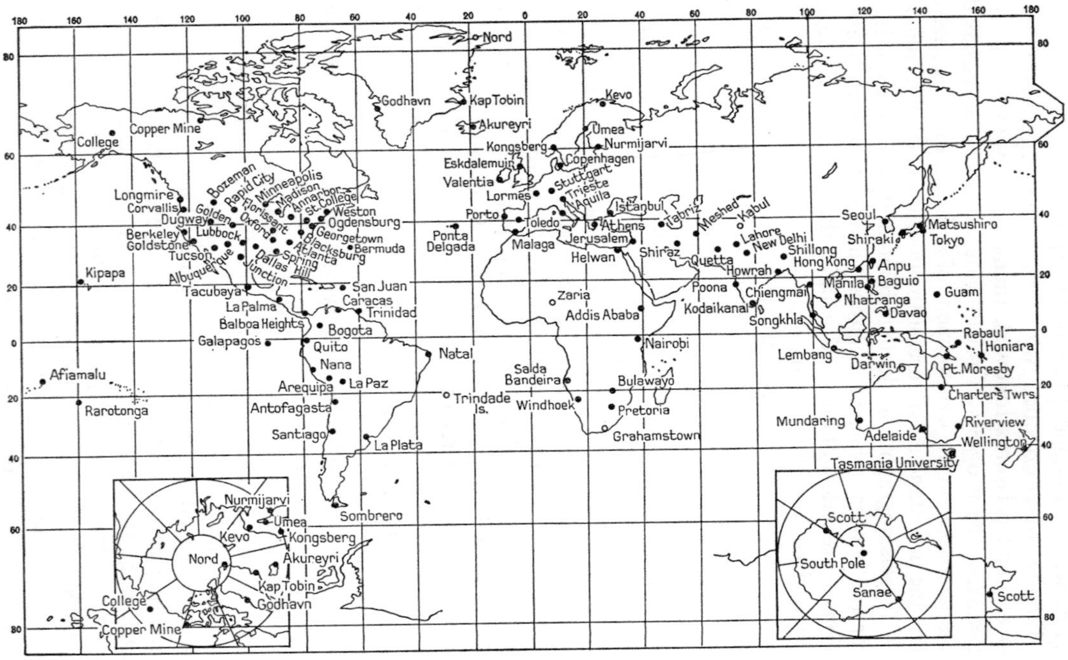

FIG 35 World-wide network of standardized seismograph stations (Courtesy of U.S. Environmental Science Services Administration, National Geophycisal Data Center)

together, give a more accurate determination of earthquake parameters than that provided by the Preliminary Determination of the U.S. Coast and Geodetic Survey. A monthly bulletin published by the Bureau gives not only the results of analysis but all the actual data used.

The second organization is the International Seismological Centre which had its headquarters at Kew Observatory for a very long time. The Centre is now based in Edinburgh and handles world wide seismic data on a massive scale, using digital computers. The data are compiled in Bulletins by the Centre and these include information from even the slowest station; they thus give the most complete coverage of seismic information available anywhere in the world.

Seismologists have always recognized the inadequacy of world coverage provided by a network of non-standard seismographs. It has also been realized that the seismic data collected need to be disseminated much faster than at present. These requirements assumed a sense of urgency when it was realized in the late 1950s that a nuclear bomb, exploded underground, behaves like an earthquake and that the instrumentation and analysis techniques of classical seismology were quite inadequate to tell quickly whether and where an event had occurred, and if is was an earthquake or an explosion. This aspect of seismology is discussed in detail in a later section. Here is it will only be said that it became very necessary to introduce some advance thinking in seismology. Progress has so far been aimed along two main lines:

(1) To set up a network of standardized world-wide stations.
(2) To set up an International Centre where data from all over the world can be sent in a form that can be directly read by a computer, and processed to produce quick, reliable bulletins. It has also been proposed that the Centre should store seismograms from all over the world, and that it should have facilities to enable microfilm copies of them to be made available to any interested seismologist, anywhere in the world.

As regards a standard world-wide network, a beginning was made during the International Geophysical Year, when the Lamont Geological Observatory of Columbia University, U.S.A. set up several stations in different parts of the world. The U.S. Coast and Geodetic Survey has, in the past few years, made a magnificent start in this direction. Drawing funds from the Vela Uniform Project of the U.S. Government, ('for intensive work in all branches of seismology, particularly with reference to detection of underground nuclear explosions'), the Survey has completed the setting up of a world network of uniform seismograph stations. Figure 35 shows the geographical distribution of the network. The Canadian Government has a network in Canada, with station spacing of the order of 500 miles, using instruments identical to those in the U.S. network. In the Soviet Union, also, similar installations are going on, even though the response characteristics of the instruments are different from the U.S. and Canadian stations. Data from the U.S. network and Canadian stations in microfilm are available, at a nominal price, to any interested seismologist.

UNESCO activities

In recent years, UNESCO has been taking a keen interest in the human aspects of seismology. The aim is to make use of our current seismological knowledge to reduce damage and loss of life due to earthquakes, where such knowledge has not been fully utilized due to economic and other reasons. The terrible destruction in the wake of the Agadir and Chilean earthquakes of 1960 made it painfully evident that countries susceptible to the worst earthquakes were doing very little about it. Accordingly, UNESCO undertook a survey of existing seismological know how in four regions of the world: Southeast Asia, South America, Mediterranean countries and Africa. A UNESCO mission consisting of seismologists visited various countries in these regions, held discussions with scientists, engineers, architects and political leaders and saw existing seismological installations. After completing the survey, the missions met and produced a report outlining clearly the steps to be taken to bring seismology to the aid of the common man.

However, for the recommendations to be implemented, governmental and economic barriers have to be crossed and every country which is directly concerned should take an active interest. One immediate result of the UNESCO activities was the formation of the International Seismological Institute in Tokyo to train world seismologists in the engineering of earthquake resistant structures and other aspects of seismological studies.

3.6 Miscellaneous Topics

Earthquake prediction

Time and again soothsayers have come forward with warning of impending earthquakes. A scientific basis for earthquake prediction, that is to say that an earthquake will affect a specified area at a given time, seemed very distant a few years ago. However, recent developments have shown that, if adequate financial support is available, a concentrated scientific effort can lead the way to effective prediction techniques in as short a period as ten years. This drastic optimism in a difficult science is based on recent work in Japan and the United States of America.

Careful measurements of relative movements between the two sides of the San Andreas and other faults in California, using very precise geodetic techniques, show that displacements are occurring currently almost along the whole length of the fault system and that these are comparable to movement rates in the geological past. The displacement rates, however, seem to differ from one part of the fault to the other. In some areas like Hollister, California, the displacements occur in 'creep episodes' of several days' to weeks' duration. Geodetic measurements made along 500 miles of California faults by the California Department of Water Resources show that in many instances 'earthquakes of magnitude 4·5 to 6 were often preceded by anomalous fault movements.' In Japan, anomalous surface and sea level changes had been reported prior to the Niigata earthquake of 1964. Laboratory studies indicate that electrical resistivity of rocks undergoes large changes prior to and during brittle fracture. F. D. Stacey has shown that a change of magnetization in rocks associated with strain build up, can be measured using an array of proton precession magnetometers. The phenomenon is termed piezomagnetism. Piezomagnetic anomalies have been observed prior to some earthquakes. Considerable advance has been made by the U.S. Geological Survey in very precisely locating micro-earthquakes in parts of the San Andreas fault. It is also becoming clear that, in some segments of the fault, strain energy is being released in the form of fault creep and small to moderate earthquakes, whereas in other segments the strain energy is released in the form of small earthquakes. Lastly, it is now becoming quite evident that human activities, like injecting fluids into the ground, filling large reservoirs, and exploding large nuclear devices in tectonically sensitive areas, can cause earthquakes in the vicinity of the disturbance. This phenomenon may lead the way to better understanding of earthquake forces and possibly to their control.

Another aspect of earthquake prediction is concerned with statistical analysis of the problem. As the occurrence of earthquakes in space and time has the nature of a statistical sample population, the data gleaned in the past can be used to determine the probability of future happenings. However, this type of study enables us to make only certain general types of pronunciations:

(1) We can say that all major earthquakes will most probably be confined to the earthquake belts of the world. We can also make statements like 'England is less likely to have earthquakes than Italy or Japan'.

(2) It can be said that a major earthquake in one region will have associated aftershocks. Also an earthquake in a given region increases the probability of earthquake occurrence in that region for some time.

(3) There is some correlation between foreshocks and large earthquakes. If frequent small tremors are experienced in a place, there is some chance that they herald a large earthquake.

A problem, closely associated with earthquake prediction, is the possibility of triggering an earthquake by forces which are relatively small and independent of the main strain-producing force. Tidal effects, temperature, abnormal pressure changes, sunspot activity, etc., can provide such 'last straw' forces. The obvious way to discover such forces's to look for correlations and repetitive patterns in earthquake occurrence. Several investigations have been made to detect these phenomena, if they exist, but no significant result has emerged.

Man-made earthquakes

Three very significant series of events which have happened since 1960, may revolutionize our understanding of earthquakes and may lead the way to the eventual control of earthquakes.

The first, and most clearly understood group of phenomena, are the swarms of earthquakes associated with fluid injection into the ground at the Rocky Mountain Arsenal in Colorado, U.S.A. In 1962 the U.S. Army started pumping waste fluids from the Arsenal into a deep well in the vicinity. Soon afterwards, the seismograph stations in the locality began recording swarms of earthquakes. Many of the earthquakes were located within 10 kms of the disposal well. The pumping was stopped in February 1966, but the earthquakes continued. In fact in 1967 there were three large earthquakes of magnitude 5 and over. Historically, the area is considered to be seismically quiet. Hence is appeared that the fluid injection into the basement had somehow released giant tectonic forces. Healy, et al (*Science* Vol. 161, pp. 1301–1310, 27 September 1968) have worked out the theory of how such forces can be triggered by fluid injection. A simplified picture of the process is as follows: A quiescent fault under the well is held together by tectonic and gravitational forces, which are quite strong enough to prevent an earthquake caused by movements of the the fault. The injection of fluid under high pressure, however, provides a hydrostatic pressure (pore pressure) which neutralizes part of the tectonic and gravitational forces, thus initiating fault movement. The lubricating effect of the fluid may also be causing a reduction in frictional forces. The 'Denver earthquakes' as the swarm is called have posed a new question, converse to the phenomenon which caused them. Can removal of liquid from under a fault reduce the likelihood of a destructive earthquake? We will have to wait for a carefully designed experiment for an answer.

The second significant event was the Koyna earthquake in India. This disastrous earthquake had a magnitude of over 7·0 and occurred in the stable Peninsular Shield of India, almost under a large newly filled dam with a capacity of 2×10^9 cubic meters and a maximum depth of about 70 meters. This region was considered to be seismically stable and had no seismic history. Since the dam reservoir started filling in 1962, numerous tremors occurred in the Koyna reservoir area. The frequency of the tremors decreased from 1963 to 1966 but afterwards began increasing again, culminating in two large shocks in September. The big earthquake occurred on December 11, 1967. It killed several people and caused severe damage in the township of Koynanager. Dams have been known to cause earthquakes. For example, the Hoover Dam in the U.S.A and Kremesta Dam in Greece have generated earthquakes in their vicinity. However, the Koyna earthquake is the most severe one of the type. It is very hard to determine the mechanism of earthquakes caused by reservoirs, that is, if it is caused by the reservoir at all. We can only speculate that possibly the elastic loading of the rocks by the water in the reservoir, combined with the hydrostatic pressure of the seeping fluid, as in the Denver earthquakes, could cause release of earthquake energy. The Koyna earthquake forms another link in the understanding of earthquakes causing forces.

Since man began testing nuclear bombs, and as their size became larger and larger, the question has often been asked, 'Can nuclear explosions cause earthquakes?' This question is now receiving serious attention, at least in the U.S.A. The following quotation from the abstract of a paper presented by Healy and Hamilton (U.S. Geological Survey) at the American Geophysical Union annual meeting, 1969, and published in EOS, Vol. 50, No. 4, April 1969 throws light on this matter.

'*Seismic activity following the "Benham" nuclear explosion*

The "Benham" nuclear explosion ($M = 6\frac{1}{2}$) and the seismic activity which followed it were monitored by 27 seismograph stations installed by the U.S. Geological Survey. As expected, many of the post-shot events occurred in the immediate vicinity of the explosion and are therefore attributed to cavity collapse. However, a large number of events occurred as far as 10 km from ground zero and appear to arise from movement along previously recognized fractures. The pattern of seismic activity changed with time, indicating a complex time-history of fracturing. Focal depths computed for the shocks range from the surface to about 5 km.'

Crustal studies using explosions

We know now the broad structure of the earth's crust is revealed by earthquake studies. A considerable amount of work on obtaining more details of the structure in specific regions, has been done using small explosions Reflection and refraction shooting techniques (see below), similar to oil exploration methods are used in the study.

Valuable contributions to our knowledge of the oceanic crust have been made by exploration work in the oceans. Much the same procedure as on land is used, except that, instead of geophones, the ship lowers a string of hydrophones. A second boat fires small charges of TNT. The first ship 'listens' while the second, starting from around eighty miles away, approaches the listening ship, firing shots at regular intervals. The shot instant is transmitted from the shooting ship by radio, by picking up the explosion signal using a hydrophone.

Oil prospecting

The techniques of earthquake seismology have been used successfully for locating oil deposits. As such deposits occur in certain types of rock formations, detected by their shape and density characteristics, it is required to plot successfully the subsurface structure in oil-bearing regions. After identifying suitable areas by geological and gravity surveys, the oil prospector proceeds to conduct seismic surveys. The seismic waves are generated by detonation of a few sticks of dynamite, and the waves travelling through various underlying layers are recorded by an array of seismographs of special design. These instruments are called geophones and are designed to have high magnification for waves with short periods because the explosion waves are much higher in frequency than those encountered in earthquake seismology.

In *refraction shooting*, the geophones are arranged in a straight line from the shot point, so that a time-distance plot (akin to a travel-time curve) can be obtained for the first recorded wave. Ideally, this plot can be split up into a series of straight line segments, the slope of each segment corresponding to particular layer velocities. Thus, the refraction shooting gives velocity distribution with depth.

The next step is to find out the actual layer depths. For this purpose the *reflection shooting* technique is employed. Geophones over a wide range of distances from the shot point are made to record simultaneously one below the other on the same sheet of paper so that reflections from deep levels can be identified on each record. From the reflection times the layer depths are computed.

Oil exploration is a highly developed technology and very sophisticated instrumentation and techniques are used to obtain the seismograms, to analyse them, and to interpret the results.

Earthquake engineering

We have already seen how one of the largest effects of earthquakes is the destruction to buildings and property. Hence it is of great importance to design buildings and other structures so that they can withstand earthquake forces. Such forces are very different from those encountered in normal civil engineering practice. As the ground moves, very complex forces, in all dimensions, are set up throughout the building. Pillars designed to carry vertical loads are suddenly subjected to horizontal forces. Roofs and superstructure find themselves no longer supported by walls which move away from them. Thus in earthquake resistant design, the engineer has to take into account the horizontal and vertical dynamic forces of an earthquake in addition to the normal static equilibrium between the forces. Naturally it is very expensive to build a completely earthquake proof building. Hence the usual convention in earthquake engineering is to design structures such that the total damage does not exceed ten per cent of the cost, and such that the damage is confined to repairable parts of the structure.

The main problems for the earthquake engineer are the dynamic forces associated with an earthquake and how to compute the response of his structures to such forces. The first thing to know is the frequency and amplitude of ground motion and the acceleration associated with it. The conventional, high magnification seismographs are of no use for providing this information as they are designed to make measurements at some distance from the epicentre. Hence the engineer uses *strong motion seismographs* which are specially designed to measure violent movements. In the U.S.A., particularly California, and in Japan large networks of strong motion seismographs are in operation.

The American instrument, designed by the U.S. Coast and Geodetic Survey is an *accelerometer* and records ground acceleration on continuously moving photographic paper. There are simpler and cheaper versions of strong motion seismographs which give only the maximum acceleration recorded during an earthquake. In one such instrument, called the seismoscope, a pendulum makes tracings on a static smoked glass and, from this record, maximum accelerations in any direction can be read out.

Since earthquakes force are oscillatory functions, the vibration characteristics of the structures must also be known before one can compute the resultant effect on the structures. Maximum response occurs when the exciting forces are of the same frequency as the natural frequency of the structure. Once a building has been constructed, strong motion seismographs can be operated in it to study its dynamic response to earthquakes. Alternatively, buildings, especially tall ones, can be subjected to artificial vibrations using a variable-speed electrical machine, and their vibration characteristics studied. But in many cases it is necessary to know at the design stage what the dynamic response of a structure is going to be. This estimation is very difficult using theoretical methods. One way out is to make scale models and measure their response on *shaking tables* (variable-frequency moving platforms used in the calibration of seismographs). Another method is to make an electrical analogue of the structure and to analyse the network using currents instead of forces. Electronic computers are extensively used in such analogue studies.

A third unknown for the engineer are the foundation characteristics of the area where the structure is to be built. To determine this, he must extrapolate from available strong motion information about similar terrain, and to apply this to his own site. Thus the response of various types of ground to earthquake forces must be known.

In most earthquake regions, building codes are prescribed by law. However, in enforcing the law, the agency responsible will not only have to supervise design characteristics but also make sure that the right materials are employed. Naturally, it is very difficult to enforce such laws, especially in economically underdeveloped countries. Experience in California and Japan shows, however, that proper adherence to building codes can reduce considerably earthquake damage to man-made structures. The UNESCO missions did their best to study existing building codes in the countries they visited, to assess how far they were enforced, and to suggest improvements wherever necessary.

Summarizing one can say that earthquake engineering and associated problems form a very important practical aspect of seismology. Progress is being made, but much remains to be done.

Normal mode oscillations of the Earth

If the Earth were given a gigantic blow, it should, being an elastic body, contract and expand alternately, and execute progressively damped oscillations until all the energy has been dissipated. The theoretical possibility of such normal mode oscillations for the case of a perfectly elastic sphere was considered by Poisson, as early as 1829. Later work by Lamb showed that, not only can an elastic spherical body oscillate by contracting and expanding along radial directions, but it can also oscillate in such a way that spherical surfaces, concentric with the centre can rotate relative to each other. The first type of oscillation, involving dilatational motion, is called *spheroidal*, and the second, involving shear, is termed *torsional* or *toroidal*.

With the Earth, as with all bodies undergoing oscillations, there are, in addition to the fundamental frequency, a series of higher modes. In 1911 Love computed that, if the Earth were a uniform elastic body, the spheroidal oscillations should have a fundamental period of sixty minutes.

For a long time studies of normal mode oscillations of the Earth were confined to theoretical seismology as the instruments were not sensitive enough to pick up the actual oscillations. However, in 1952, following the Kamchatka earthquake of November 4, Benioff recorded waves of 57 minute period on his strain seismograph. This stimulated more theoretical work on the subject and experimental search for more conclusive data to prove the existence of oscillations of the Earth.

Several computations of normal-mode periods were made, allowing for the known inhomogeneities of the Earth. The mathematics involved in these computations is very complex and the calculations themselves long and tedious. It is doubtful whether extensive theoretical work in the field could have been possible without the aid of electronic computers. Results showed that the Earth, as a whole, can oscillate at various periods, the highest being around the 57 minutes found by Benioff and very close to Love's sixty minutes. It was also postulated that the Earth's core itself can have a fundamental spheroidal mode period of about 100 minutes. Since the computed periods were dependent on the density and elastic structure of the earth, here was a chance to get a good check on the various Earth models. However, for seven years after the Kamchatka earthquake, the search for normal mode oscillations on seismograms met with no success. Then, in 1960, at the meeting of the International Association of Seismology and Physics of the Earth's Interior, in Helsinki, the most dramatic session witnessed in seismology in recent years occurred. In the words of Professor Bullen, the veteran seismologist. 'Press announced that Benioff had once again observed long-period waves, this time from the Chilean earthquakes of 1960, May 22. Thereupon Slichter announced that his group had observed similar long-period waves, not on a seismograph, but on a La Coste-Romberg tidal gravity meter. Comparison showed that a number of periods of the two groups were in good agreement, notably periods of about 54, 35·5, 25·8, 20, 13·5, 11·8 and 8·4 minutes, but that certain of Benioff's periods were missing in Slichter's results. Pekeris, who was also present at the session, studied the missing periods and announced that these periods corresponded to his calculations for torsional oscillations, and further that these would not be recorded on a gravity meter. The two sets of results were thus shown to be in remarkable accord and all doubt as to the genuine recording of free long-period oscillations of the Earth was removed.' Subsequently, it was revealed that long-period oscillations were recorded on pendulum seismographs as well.

The comparison of observed periods with theoretical values obtained for various models gave a valuable check on how good a particular model was. Many important studies on the spherical symmetry of the Earth, nature of the inner core, etc. were made using the normal mode results. The details are beyond the scope of this essay.

The normal mode oscillations are the most important discovery in seismology so far in the latter half of the twentieth century. It proves what can be achieved with superior instrumentation and advanced techniques.

Microseisms

At the beginning of this essay it was stated that, even when there are no earthquakes, the Earth is never completely at rest. Given superhuman eyes, one might see that the Earth is as 'restless' as the ocean surface. The general term for the continuous, omnipresent background seismic noise is 'microseisms' or 'earth noise'.

Some of the obvious sources of microseisms are man made or *cultural*, like factories, traffic, trains, aircraft, pumps, etc., some *meteorological*, like wind blowing on tall structures and trees and atmospheric turbulence setting loose soil into vibrations.

In general, these microseisms lie in the frequency range of about one to several Hertz. At the other end of the scale are slow earth movements caused by atmospheric loading on the Earth's surface, temperature effects, oceanic tides, interplanetary forces and orogeny (mountain building movements). In between these two, in the frequency band from one to probably 1/30 Hz, occur some very interesting microseisms caused primarily by ocean waves. These are called *storm microseisms* or *oceanic microseisms*.

Even the early, crude seismographs showed large, regular earth movements with a predominant period around six seconds at different parts of the world. Careful study showed that the earth moved on an average only about a micron (1/1,000 mm), but that, at times, the amplitude of the waves increased by a factor of as much as fifty and remained high for several days. The average amplitude in winter was much higher than in summer and more 'microseismic storms' were observed in winter. These observations pointed to bad weather as a probable cause of microseisms, and an early theory attributed them to surf beating on coast. However, it was soon observed that high microseismic activity was sometimes recorded at seismic stations, even before ocean waves were appreciable in the neighbouring coastal regions. As more data accumulated, there was sufficient evidence to believe that, in certain cases, in addition to generation at the coast, microseisms could be generated directly underneath storms in the deep ocean. Since seismic waves associated with microseisms travel at speeds about 200 times faster than storms and ocean waves, they reached land stations much before their originators reach the coast. This idea gave birth to a valuable use of microseisms in meteorology, to track the formation, intensification and movement of storms. During the Second World War, several attempts were made, particularly in Great Britain and the United States, to use microseisms for storm forecasting. Although no one was quite sure how energy from a surface storm was communicated through the deep water layer to the ocean bottom to generate seismic waves, several empirical techniques were devised to use microseisms for storm tracking and for predicting wave conditions on distant beaches. However, due to the empirical approach and to several unpredictable features of microseisms themselves, no systematic success crowned the efforts, even though valuable data were accumulated.

After the war, microseisms still retained a certain amount of academic interest and curiosity. Theoretical work, notably by Longuet-Higgins in England, showed that microseisms were generated by the interference of

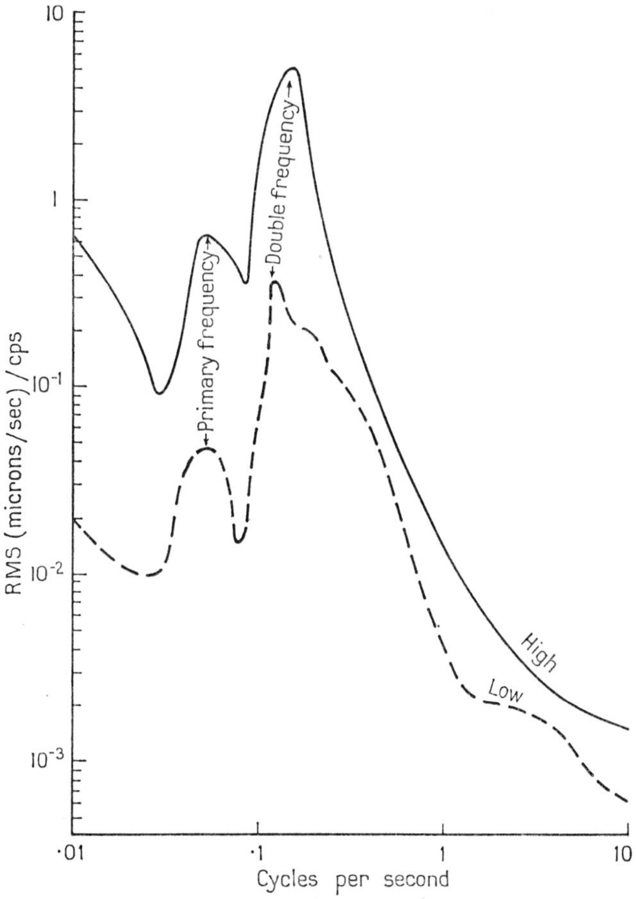

FIG 36 Haubrich's average spectrum of earth-noise showing the *primary frequency* and *double frequency* peaks. The vertical scale is in terms of energy associated with the microseisms

criss-crossing ocean waves in the vicinity of storms, and incident and reflected waves in coastal areas. Such interference, similar to the familiar standing wave phenomena, set the water layers of the ocean into normal mode oscillation, as if they were huge organ pipes.

Very precise instrumentation and analysis techniques, in England and later in U.S.A. have revealed many fundamental properties of microseisms associated with storms and ocean waves. For example, it is known that microseisms are predominantly surface waves. Using high magnification seismographs and digital computer analysis, Haubrich in California detected two types of oceanic microseisms. The first is called a *Primary Frequency* microseism and has the same frequency as the ocean waves which generate them; primary frequency microseisms are caused by direct loading of the earth by ocean waves piling up water near the coast. The second type is called a *Double Frequency* microseism and is caused by interfering ocean waves over deep ocean, as already described. These microseisms have a predominant period of about seven seconds. Figure 36 shows a spectrum of earth noise showing the *P.F.* and *D.F.* peaks.

On average, oceanic microseisms move the Earth only by a fraction of a micron to a few microns. The higher frequency *cultural* and *meteorological* microseisms vary sharply from place to place, but move the Earth by only a few thousandths of a micron. However, even these very minute movements can mask the waves from small seismic events. A new requirement in seismology (see next section) is to detect very minute seismic signals produced by small nuclear explosions. To be sure that these signals can be recognized, even in the presence of background noise, it is necessary to find 'quiet' sites for installing detection instrumentation. Optimum signal/noise processing techniques, as used in other fields like radio engineering and radio astronomy, will also help. Hence, the study of microseisms, their origins and properties, occupies a very important place in seismology today.

Nuclear seismology

The first atom bomb was exploded in New Mexico on 16th July, 1945. It was detonated from the top of a tower 100 feet high. The explosion, though above ground, produced detectable seismic waves which were recorded and studied. Since then, the seismic effects of nuclear explosions have produced an unprecedented development in seismology. For the seismologist, the nuclear bomb explosion provides a precisely known source, thus eliminating several of the unknowns associated with earthquake research. However, the rapid development in seismology caused by nuclear explosions is not the result of our desire to exploit the potentialities of the bomb to study accurately the structure of the Earth. The reasons are more political, since the only way to detect an underground explosion at a distance from the source is from the seismic waves it produces. Hence, it has become an international necessity to understand all aspects of the generation of seismic waves from nuclear explosions, and also to distinguish these waves from earthquake waves. Adequate instrumentation for the *detection* of such explosions, and comprehensive techniques to process the data for *identification* are a necessary prelude to any agreement on complete cessation of nuclear testing. This requirement has channelled large funds into all aspects of seismological studies. The multi-million dollar U.S. budget to support the VELA UNIFORM programme is an example.

A nuclear explosion of yield 1 kiloton (equivalent of 1,000 tons of TNT) produces energy of the order to 4×10^{19} ergs. A 20 kiloton explosion has a total energy release of about 10^{21} ergs. The large 100 megaton Soviet explosion of October 1961 released almost as much energy as a large earthquake. However, luckily, not all of this energy is converted into seismic effects. An explosion on the earth's surface couples only 1/10 000 of its total energy to the ground. The coupling factor is termed *seismic efficiency* and is greater, the deeper the explosion. The seismic efficiency is highest for underwater explosions and is of the order of 1/25 for a shot 500 metres underwater.

As already stated, the only detectable effect from an underground nuclear explosion at a distance from the source is the seismic effect. The seismic signal from an explosion is practically indistinguishable from earthquake signals, but we do know that the explosion source is small (relative to an earthquake source) and omni-directional, that is, it releases energy in all directions, whereas an earthquake, being a slide phenomena, releases energy in certain directions only. Hence there is bound to be some difference between the seismograms from the two types of source. The aim of nuclear seismology is to develop techniques to detect small seismic events and to analyse the seismograms with a view to picking out even the minute differences between explosions and earthquakes.

Detection problem

The first problem is the detection of small seismic events. A 20 kiloton explosion has a magnitude of only about 5, and a one kiloton explosion is really small, with magnitude of about 3·7. The efficiency of a nuclear detection system depends directly on the smallest magnitude it can detect as far away from the source as possible. This requirement is something new in seismology, as classical seismology got all the data it wanted from a few large earthquakes and was not concerned with small earthquakes at large distances. The first necessity, therefore, is to develop very sensitive seismometers and high fidelity recording systems. However, the limit of usefulness of a seismograph is soon reached due to the omnipresent background noise of the earth. As the gain of the seismograph is increased, it only amplifies weaker microseisms, other small seismic signals being completely engulfed in them. Fortunately, the amplitude of microseisms varies from place to place. To be of maximum use, a nuclear detection system should be set up in the *quietest* spot, seismologically speaking, so that the optimum signal/noise ratio can be obtained. The necessity to consider noise and signal/noise ratio brings seismology onto a par with other fields of science like communication, both acoustic and radio, and radio astronomy.

Given a *quiet* spot, the next step is to arrive at a pattern of seismographs and to process the data from them in such a way as to improve signal to noise ratio. Such patterns of instruments are a familiar feature in communication and radio astronomy, where arrays of antennae are used to enhance signals in particular directions. Hence arrays of seismographs are used to detect small seismic events.

'Noise' in nuclear detection work, is in the frequency region where the *P* and *S* waves usually occur. An accepted spectral range is from 0·5 – 3 Hz. Since the bulk of seismic noise in this range is locally generated 'cultural noise,' the seismic network should be as far as possible from sources of such noise—cities, factories, railroads, etc. Secondly, as the ocean is also a noise source, mid-continental sites are the quietest. Lastly, it has been found to be advantageous to bury the seismographs in boreholes several hundred feet deep to reduce the surface seismic noise. At one time it was conjectured that the deep ocean basins might be very quiet and ideal locations for detection systems. However, recent measurements using deep ocean seismographs (see later section) have provided little support for the 'silent deeps' theory.

The details of the use of arrays of seismometers to improve signal to noise ratio are beyond the scope of this essay, but it is worth noting that the improvement rises with the number of instruments used for a single operation known as steered summing. Thus, if the signal and noise amplitudes are the same at a particular place, it is not possible to identify the signal at all on a record from a single seismograph; but, by summing outputs from ten seismometers distributed in a certain way, one can obtain a signal whose amplitude is three times that of the noise. An experimental seismic array for detecting nuclear explosions in distant parts of the world is operated by the United Kingdom Atomic Energy Authority at Eskdalemuir in Scotland. The set-up consists of 22 seismometers, arranged in two perpendicular intersecting arms of eleven each, at right angles to each other. The electrical signals from individual seismometers, suitably preamplified, are telemetered to the central laboratory on the site and recorded continuously on magnetic tape. The finished tape reels are processed at a laboratory, where operations such as *steered sum* are performed on the data using special computers.

Identification

After an event has been clearly detected above background noise, it is required to say whether it is an earthquake or an explosion. The problem is very complex and several years of research have produced no hard and fast criteria that can be applied. The following are some of the characteristics of earthquakes and explosions which, when taken together, can establish an event as 'suspicious':

(1) One of the first methods suggested to discriminate between an earthquake and an explosion is the direction of movement of the earth when the first signal arrives. The majority of earthquakes are generated by a shearing movement of rock blocks, and the seismic waves generated from such motions can be expected to be compressional in certain directions and rarefactional in others. Explosions, on the other hand, produce compressions in all directions. Thus, if samples of seismograms from widely distributed stations show the first signal to be a compression in every case, the event can be classified as suspicious. Unfortunately, this criterion is very difficult to apply in practice as the first motion in a seismogram is usually very small, and it is difficult to estimate accurately the direction of movement.
(2) With most earthquakes, the shear mechanism at the focus can be expected to produce *S*-waves as prominently as *P*-waves. Explosions, do not generate *S*-waves at the source, but, on seismograms, show prominent shear waves produced during *P*-wave travel. However, it is reasonable to hope that there would be more shear waves from earthquakes, and the ratio of *P*- to *S*-wave in a seismogram can therefore provide an important clue as to the nature of the source. Similarly, energy partition between Rayleigh and Love waves can provide some guidance about the nature of a seismic event. Here again, there are several practical difficulties in applying the criteria, as the travel from source to recording station introduces very many complications.
(3) Nuclear explosions usually take place closer to the surface of the earth than earthquakes, so, if the depth of focus can be estimated accurately, very shallow events can be classified as suspicious. However, our current accuracy in the estimation of depth of focus is very small. An event classified as occurring 25 kms down could have happened anywhere between the surface and 25 kms deep. However, modern data processing techniques can be expected to improve the accuracy.
(4) We know fairly well the distribution of seismic zones on the surface of the Earth. Hence if an unusual event occurs in an aseismic region, it can be treated as suspicious.

There are several more *diagnostics* that can be employed to increase the accuracy of identification in a nuclear surveillance system. The above points will show that the system, and the analysis and interpretation are far from simple, and open up new frontiers in seismology.

The Geneva Committee in 1958 envisaged a world-wide network of 180 *control posts* for the detection of nuclear explosions. The recommendations of the Committee could not be implemented as the international machinery required for successful operation of the network was quite complex.

However, as a result of the deliberation, intensive national efforts in seismology are being made to develop sophisticated surveillance systems. The U.K. and U.S. efforts are well known and it can be inferred that several other countries are also engaged in similar work.

From the civilian point of view, the use of data from several explosions whose location, origin, time and yield are published, has enabled very accurate studies on crustal and upper mantle structure to be made. The data have also helped in arriving at valuable regional corrections to travel-time tables, with consequent benefit to earthquake seismology.

Seismicity of extra-terrestrial bodies

We have seen that an abundance of information on the structure of the earth has been collected by studying earthquake seismograms. The same idea can be used to obtain information regarding the seismicity, if any, and interior structure of other members of the solar system. Being our closest neighbour, the Moon has been our first target. The Ranger and later Surveyor series of experiments to land a seismometer on the moon were not successful. However, as this is being written, the seismometer planted on the moon by the APOLLO 12 astronauts is telemetering 'moonquake' data back to Earth.

What can we learn about other planets by seismic methods? Naturally we will study their internal structure, and this will not only enhance our knowledge about the planets themselves, but will also throw valuable light on their evolution in relation to earth, and on the history of the solar system. Another interesting experiment will be recording meteorite impacts, of which there are very many in outer space. Because of the absence of oceans and weather on the Moon, any microseismic activity there, will be due to solar and radio-active heating effects of the moon rocks. Thus seismic noise studies may be able to contribute to our knowledge of the thermal condition of the moon.

With a view to conducting a lunar seismic experiment, the California Institute of Technology and Columbia University, U.S.A., have been working on the design of special seismographs. The California Institute's seismographs have been used in several of the *Ranger* series of space vehicle launchings, and this fact alone demonstrates the transition of seismology from smoked-paper records to computer analysed traces in just over half a century.

The requirement was for an instrument, sufficiently light and compact to be carried in a space vehicle, which could operate at extremes of temperatures and which could withstand severe acceleration during take off, vibrations during transit, and impact during landing. It also had to have electronic amplifiers, a telemetry system and sufficient power to operate for several weeks on the lunar surface.

A seismometer with a natural frequency of one second, about 0·7 critical damping and a velocity (moving magnet) transducer was designed for the purpose. The instrument was 130 mm high, 110 mm in diameter, and weighed only 3·45 kg. It was completely sealed and filled with an organic fluid. The instrument was tested by 600 foot drops from a helicopter on various types of ground, including runways. The seismometer was housed in a survival sphere with electronic equipment, power supply, temperature control devices, etc. This sphere, together with its retro-rocket was carried on the launch vehicle. Before landing on the moon, the capsule with retro-rocket was to separate, the retro-rocket was to fire, and slow the velocity of the capsule. After burn out, the retro-rocket was to separate, letting the capsule fall free on the lunar surface.

After landing, the instrument was to align itself with the local vertical and transmit amplified signals to the earth by radio. At the receiving end the signals were to be decoded and written as visible seismic records and also on magnetic tapes for computer analysis.

Ocean bottom seismology

A logical outcome of modern developments in seismology is the attempt to plant seismometers on the ocean bottom. Since more than two-thirds of the Earth's surface is covered by the oceans, it is obvious that a complete global seismic network should have at least twice as many instruments on the ocean bottom as on land. In addition, the problem of microseism generation by ocean storms, which baffled seismologists for nearly a century, warranted some direct ocean bottom measurements. It was also considered probable that, under the deep oceans, the seismic noise would be very low and hence the environment should be ideal for nuclear surveillance stations. In any case, it was finances from the bomb detection projects and advancement in oceanographic instrumentation and space seismology, that made it possible to visualize planting seismographs on the ocean bottom.

Three possible ways have been tried for making ocean bottom seismic measurements:

(i) A seismograph (water tight and pressure proof) is lowered from a ship to the sea-floor and left for a short length of time. The seismic signals from the instrument are telemetred to the ship via cable. However, this technique has severe limitations in the sense that the heaving of the ship and the dragging of connecting cable tend to communicate spurious movements to the instrument packet. Monokhov of U.S.S.R. has used the technique fairly successfully and has made measurements on the floor of the Black Sea and the Baltic. He used a single vertical component seismograph, mounted with self-aligning bearings in a heavy steel sphere, which was lowered into the ocean by a steel cable. The sphere also contained a galvanometer and recorder.

(ii) The seismograph package is dropped from the ship and falls freely to the ocean bottom. The data is telemetred to the ship by an acoustic link. Such a system was used by the Lamont Geological Observatory of Columbia University. The instrument consisted of a geophone, with an electronic package which sent a frequency modulated acoustic signal to the surface ship. The signal was picked up by the ship's echo sounder.

(iii) The seismograph package with associated electronics and recording system is dropped by free fall so as to settle on the ocean bottom. It records for a pre-set interval of time and refloats to the surface to be picked up by the ship. The Institute of Geophysics of the University of California, where oceanic microseisms are studied, uses this type of package. Three seismometers, of natural frequency 1 Hz, and of the same type as the California Institute of Technology lunar seismograph, are mounted inside a hollow aluminium sphere, which can be

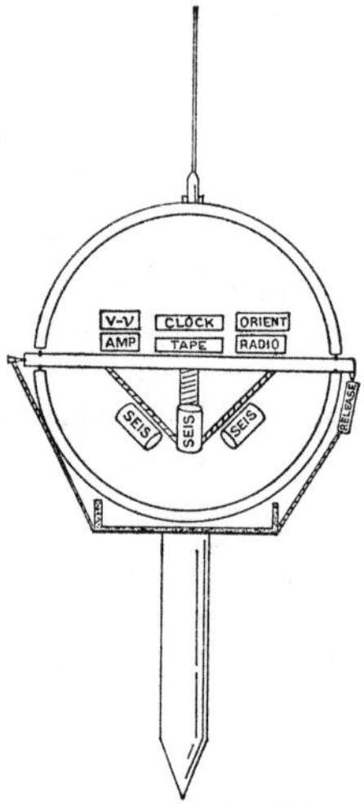

FIG 37 University of California's ocean bottom seismograph package. (By courtesy of H. Bradner and *Science,* Vol. 146, pp. 208–216, 9 October 1964. Copyright 1964 by the American Association for the Advancement of Science)

rigidly attached to a seventy kilogram lead anchor. The sphere with anchor is dropped overboard. After a preset recording time, the anchor gets detached from the sphere by a slowly dissolving magnesium link. The sphere floats to the surface and actuates a radio transmitter which sends signals for the ship to come and pick it up. The sphere contains electronics for the seismograph, a magnetic tape recorder and power supplies. Figure 37 illustrates the University of California's tape recording seismograph (figure 9, Bradner).

Extensive ocean bottom work is also done by the Texas Instruments Company of the U.S.A. They have used a 'tethered seismograph system.' The instrument package consists of three component seismographs, and it is lowered by a cable, anchor and chain system, so that it falls freely through the last six metres of its travel, thus penetrating into the ocean bottom to get a good coupling. The 300 metre length of chain connecting the package to the 350 kg anchor effectively decouples the instrument from cable vibrations. Figure 38 shows the principle of the installation of the Texas Instruments Company ocean bottom seismograph. The signal is transmitted to the ship by cable and recorded on strip chart and magnetic tape recorders.

Ocean bottom seismology, just like space seismology, is still in its initial instrumentation stages and it is too early to say much about the nature of seismic signals recorded. Only one thing seems to be clear: namely, that ocean bottom seismic noise is, in general, several times higher than on land.

Continental drift, ocean floor spreading and the new global tectonics

The idea that all the continents of the Earth were joined together at one time and have drifted apart to their present positions is a very old one. By the end of the 19th century and the beginning of the 20th century, scientific evidence for this hypothesis began accumulating. Thus, L. Wegner of Germany presented remarkable geological and palaeontological evidence showing similarities on both sides of the Atlantic Ocean and proposed that the continents were probably joined together about 200 million years ago. During the first 60 years of this century, 'continental drift' was a subject of hot debate between proponents and opponents of the theory. However, during the late fifties and early sixties, a remarkable set of world-wide geologic and geophysical observations have led to the discovery of a new type of phenomenon, closely related to continental drift. This is the theory of ocean-floor spreading and it is now causing a revolution in Earth science. According to this theory, the ocean floor, consisting of the thin crust and part of the mantle and termed the 'lithosphere', is being generated at the oceanic ridges and moves like a conveyor belt to disappear under the ocean trench-island arc systems. The crucial measurements for this theory are based on palaeomagnetic studies and are very precise. It is known from magnetic studies of lava flows in volcanoes and deep-sea sediments that magnetic reversals have occurred synchronously all over the world during geological history. Nine magnetic reversals seem to have occurred during the past 4 million years. According to the ocean-floor spreading hypothesis, the oceanic ridges are created by rising currents of material which spread out on either side to generate new ocean floor. As the rising material cools, it 'freezes' within it the instantaneous magnetic history, and carries it along the ocean bottom. Thus there tend to be magnetic anomalies symmetrically

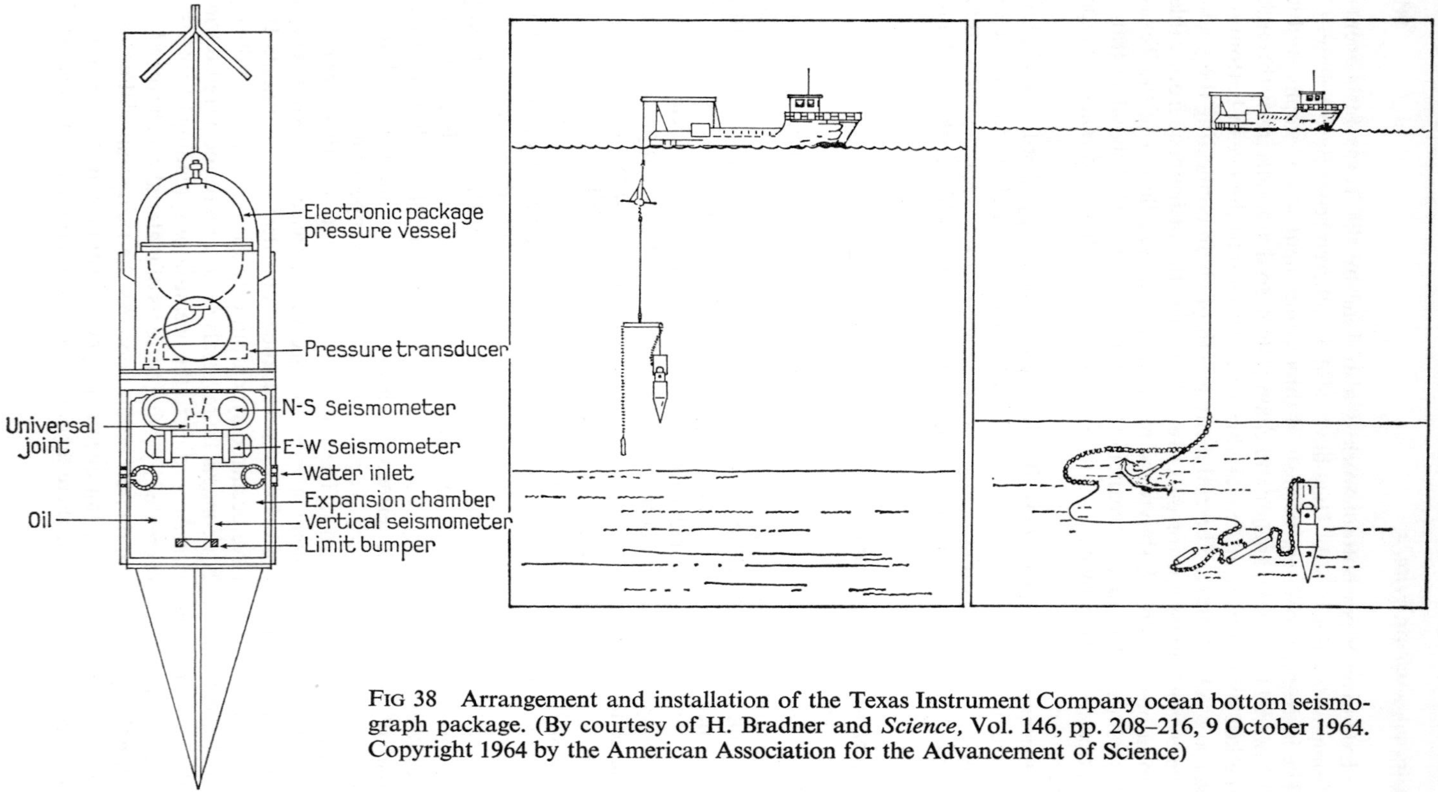

FIG 38 Arrangement and installation of the Texas Instrument Company ocean bottom seismograph package. (By courtesy of H. Bradner and *Science*, Vol. 146, pp. 208–216, 9 October 1964. Copyright 1964 by the American Association for the Advancement of Science)

on either side of oceanic ridge systems. The lithosphere is absorbed under the oceanic trench-island arc system. (Figure 39 illustrates the mechanism schematically). The ocean-floor spreading rate is of the order of a few centimetres per year. The driving forces for these giant conveyor belts are considered to be large convection currents in the hot plastic mantle. Though the ocean-floor theory can exist independently of the continental drift theory, it seems reasonable to speculate that both are different manifestations of the same physical phenomena. The present concept visualizes continents as being carried along with the ocean floor like 'logs frozen in ice' (Tuzo Wilson).

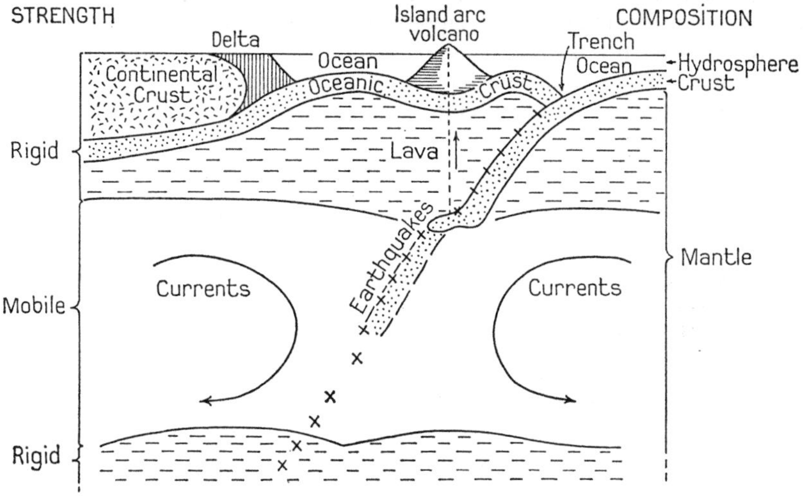

FIG 39 Reabsorption of the oceanic crust into trenches on the convex side of island arcs. (Reproduced from J. Tuzo Wilson, A Revolution in Earth Science, *Canadian Mining and Metallurgical Bulletin,* February 1968)

What has seismology contributed to this revolutionary theory of ocean-floor spreading? Earthquake studies have led to the recognition of a world-wide ocean-ridge system, the 'source' of the ocean floor. A striking feature of the ridge system is that it is characterized by offsets in a number of places. The hypothesis of the new global tectonics is that these offsets are due to relative movements between two planes of the lithosphere; they are termed 'transform faults'. Detailed studies of earthquake source mechanisms, using first-motion seismic data by Stauder, Isacks and Sykes have shown that the sense of movement is different, depending on whether the earthquakes are associated with the ridges or the transform faults. The ridge earthquakes are characterized by *dip-slip* movements, whereas the transform fault earthquakes show *strike-slip* movement indicating lateral motion. The striking phenomenon associated with island arcs is the existence of deep focus earthquakes (see section on geographical distribution of earthquakes). By very careful study of earthquakes under the Tonga-Kermadec trench, Sykes has shown that the seismic activity is confined to a well-defined narrow zone (the Benioff zone) which dips beneath the island arc at an angle of about 45°. The thickness of the seismic zone is less than 100 km. In addition, studies of focal mechanisms of both the shallow and deep shocks, and the attenuation of seismic waves from the shocks support a model of a cold thin plate plunging into the hot mantle below.

Thus, in a very short time, a new global tectonics based on continental drift, and sheets of ocean-floor spreading and moving relative to each other creating transform faults and disappearing under island arcs, has come into existence. Very precise locations of earthquakes along ocean ridges and island arcs, studies of the mechanism of such earthquakes, and measurements of velocities and amplitudes of the waves have significantly contributed to this revolution.

Further reading

Barazangi, Mauwia and James Dorman, 'World Seismicity Maps Compiled from ESSA, Coast and Geodetic Survey, Epicenter Data 1961–1967', *Bulletin of Seismological Society of America,* Vol. 59, No. 1, pp. 369–380, February 1969.

Benioff, Hugo, 'Earthquake Source Mechanisms', *Science,* Vol. 143, No. 3613, pp. 1399–1406, March 27, 1964.

Bradner, Hugh, 'Seismic Measurements on the Ocean Bottom', *Science,* Vol. 146, No. 3641, pp. 208–216, October 9, 1964.

Bullen, K. E., *Introduction to the Theory of Seismology,* Cambridge University Press, London and New York, 1965.

Eiby, G. A., *About Earthquakes,* Harper and Bros., New York, 1957.

Gamow, George, *Biography of the Earth,* A Mentor book published by the New American Co., New York, 1955.

Healy, J. H., Rubey, W. W., Griggs, D. T., and C. B., Raleigh, 'The Denver Earthquakes, *Science,* Vol. 161, pp. 1301-1310, September 27, 1968.

Hodgson, John H., *Earthquakes and Earth Structures,* Prentice Hall, Inc., New Jersey, 1964.

Howell, Benjamin F., *Introduction to Geophysics,* McGraw Hill, 1959.

Isacks, Brian, Oliver, Jack, and Lynn R., Sykes, 'Seismology and the New Global Tectonics', *Journal of Geophysical Research,* Vol. 73, No. 18, pp. 5855–5899, September 15, 1968.

Iyer, H. M., 'Earth Noise', *New Scientist,* No. 304, September 13, 1962.

Richter, Charles F., *Elementary Seismology,* W. H. Freeman and Co., San Francisco, 1958.

Strahler, Arthur N., *The Earth Sciences,* Harper and Row, 1963.

Stubbs, Peter, 'Simpler Detection of Underground Bomb Tests', *New Scientist,* No. 297, pp. 186–188, July 26, 1962.

Tazieff, Haroun, *When the Earth Trembles,* Rupert Hart-Davis Ltd., London, 1964.

Wilson, J. Tuzo, 'A Revolution in Earth Science', *Geotimes,* pp. 10–16, December 1968.

Geomagnetism

F. D. Stacey, Reader in Physics, University of Queensland

4.1 The Main Field

It has been known for many centuries that there is a magnetic field everywhere on the Earth. The strength and direction are sufficiently constant to enable mariners to determine their bearings and partly for this reason it has been charted with increasing precision. The overall magnetic pattern has a general similarity to the field which would be produced at the surface of the earth by a powerful dipole magnet or current loop situated at the centre, with its axis inclined to the geographic axis by about 11°.

With the advent of satellites, measurements of the geomagnetic field now extend outwards into space for tens of thousands of kilometres and we know that the field embraces a volume many times the size of the Earth. By contrast, direct probing towards the Earth's interior is confined to the few kilometres of the Earth's crust rendered accessible by mining and drilling operations. Nevertheless it is evident that the main part of the earth's magnetic field is of deep internal origin. Magnetic fields due to effects external to the Earth are small by comparison with the main field and are mostly transient in nature.

Repeated careful measurements over many years have shown not only that the main field is varying slowly, but that there is a systematic pattern of change which gives a vital clue to the origin of the field.

The magnetic elements and magnetic maps

Magnetic field is a vector quantity which requires the specification of three *elements* for a complete statement of its magnitude and direction at any point. There are alternative ways of choosing sets of three elements. The most usual set is horizontal component (H), vertical component (Z, sometimes represented by V) and declination (D), which is the angle between the direction of the horizontal component (the magnetic meridian or magnetic north) and the true or geographic north. An alternative set of elements is total field intensity (F), dip or inclination (I), with respect to the horizontal plane, and declination. Occasionally vector components are referred directly to geographic coordinates, north (X), east (Y) and down (Z). Data may be converted from one set of elements to another by simple trigonometrical substitutions. Thus to calculate (F, I, D) from (H, D, Z) or vice versa the following relationships are used:

$$F^2 = H^2 + Z^2 \quad \text{and} \quad \tan I = Z/H,$$

or

$$H = F \cos I \quad \text{and} \quad Z = F \sin I.$$

All of the elements have been plotted as world maps, which show iso-magnetic lines or contours joining points of equal value for each element. The one most frequently charted is declination, D, which is universally used in navigation and frequently termed 'variation' by seamen. It is measured by compasses or magnets which are suspended so that they are free to rotate in a horizontal plane and thus to align themselves with magnetic north. Except at the dip equator, where I is zero, the field is inclined to the horizontal, so that a compass does not measure the direction of the total field, but only its horizontal component.

The dipole and non-dipole fields

A revealing way of displaying the geomagnetic field on a map is to compare it with the field which would be produced by an ideally small magnet or dipole located at the Earth's centre and to plot the difference. The field of the dipole which gives the closest fit to the observed field is termed the *dipole field* and the vector difference which is obtained by substracting it from the observed field is the *non-dipole field.*

The field of an ideal dipole is shown in cross-section in Figure 40. Mathematically we may consider a dipole to consist of two magnetic poles, $+m$ and $-m$, whose physical size and separation d are infinitesimal, but whose dipole moment, $M = md$, is nevertheless finite. A uniformly magnetized sphere has a field identical to that of a dipole and the field of a short bar magnet is similar except at points close to the magnet. The field at magnetic co-latitude θ (90° minus latitude) and distance r from a dipole of moment M, as in Figure 40, may be expressed in terms of its vertical (i.e., radial) and horizontal (i.e., circumferential and polewards) components, which are the magnetic elements Z and H:

$$Z = \frac{2M}{r^3} \cos \theta,$$

$$H = \frac{M}{r^3} \sin \theta.$$

As the axis of the coordinate system has been chosen to coincide with the axis of the dipole, there is, in this case, no latitudinal component of the field, as represented by the declination when the Earth's field is referred to geographic co-ordinates.

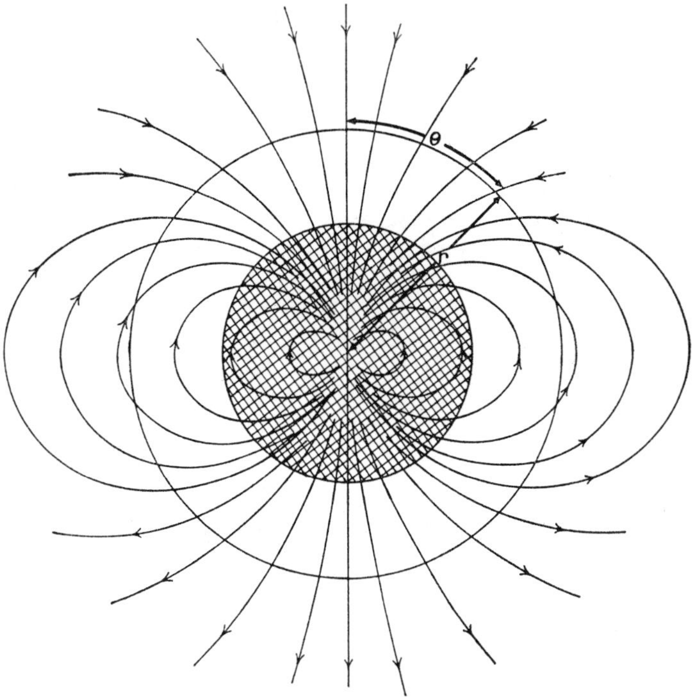

FIG 40 Lines of force in a dipole field. A dipole field is conveniently represented in polar co-ordinates, r being the distance of any point from the centre and θ its magnetic co-latitude or angle between the dipole axis and the radius to the point. The field is here represented as being that of the Earth, the outer circle being the earth's surface. The shaded inner circle is the liquid iron core within which the field is generated. The radius of the liquid core (3470 km) is slightly more than half that of the whole Earth (6370 km).

All of the important characteristics of the dipole field may be calculated from the above equations; thus total field strength is:

$$F = (Z^2 + H^2)^{\frac{1}{2}} = \frac{M}{r^3}(1 + 3\cos^2\theta)^{\frac{1}{2}}$$

and the angle of dip, I, is given by

$$\tan I = \frac{Z}{H} = 2\cos\theta.$$

The equation of any line of force is:

$$r = r_e \sin^2\theta,$$

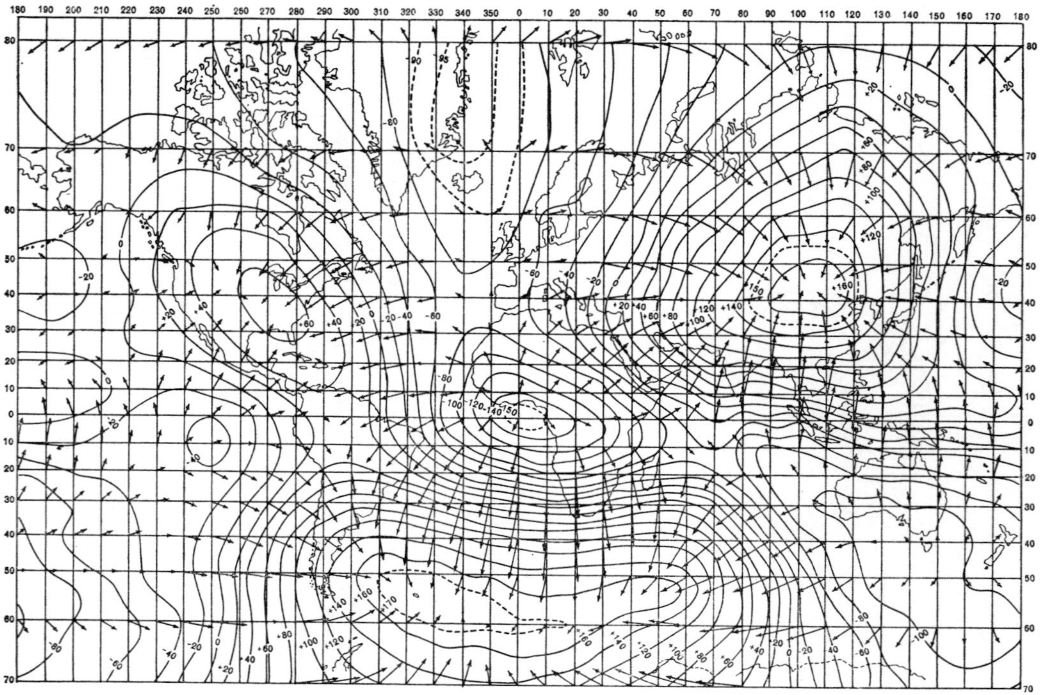

FIG 41 The non-dipole field for 1945. The vertical field intensity is contoured and the contour values are marked in units of 10^{-3} oersted (100 gammas). Arrows show the direction and magnitude of the horizontal component. A length corresponding to 10° of longitude represents 0·064 oersted. (Figure by permission of the Royal Society from E. C. Bullard and others, *Phil. Trans. Roy. Soc.*, London, A243, p. 70)

r_e being the maximum distance from the centre which the line reaches at the point where it crosses the equatorial plane. Applying these equations to the Earth, the closest fitting dipole is found to have a moment of approximately 8×10^{25} electromagnetic units, with an axis inclined at about 11° to the geographic axis. This corresponds to a horizontal field of approximately 0·3 oersted at the magnetic equator and a vertical field of approximately 0·6 oersted at each magnetic pole.

By subtracting the dipole field from the observed field, the non-dipole field as shown in Figure 41 was obtained by E. C. Bullard and others. Local variations of small extent are of course smoothed out on a map of this scale, which shows the large areas, each several thousand kilometres in extent, to or from which lines of magnetic force converge or diverge. It is not difficult to imagine such a field being produced by an arrangement of a few powerful dipoles at great depth within the earth. In fact the non-dipole field is quite well represented by eight vertical dipoles distributed around the core of the Earth, more than 2,900 km down.

A close look at the dipole field shows that superimposed upon the large scale features, which are upwards of 2,000 km in extent and are apparent in Figure 41, there are small scale features from a few kilometres up to a maximum of about 200 km in extent. However, there is a notable absence of features of intermediate sizes. This is directly indicative of the origin of the field. The large scale features have a deep origin, in the core, and are properly parts of the main field of which the dipole field is the strongest part. The small scale features are due to the rocks in the Earth's crust the uppermost 5 to 40 km, which are magnetic down to a depth of about 20 km, below which the temperature is above the Curie point of magnetite and all minerals are purely paramagnetic. The absence of features of the field of intermediate size indicates that there are no sources of the field in the mantle the 2,900 km thick solid part of the Earth between the crust and the core.

Secular variation and the westward drift of the non-dipole field

Measurements over long periods reveal steady, progressive changes in the main field. They are immediately apparent from yearly averages of the values of magnetic elements reported by all of the world's magnetic observatories. By taking period averages we eliminate from consideration the transient disturbances, which are due to interaction of the field with particle emissions from the sun, and leave only the slow variations which are of internal origin and are known as the *secular variation.*

The discovery of the secular variation was made in 1634 by Gellibrand who had, by that time, access to several decades of observations of the magnetic declination at London. When coupled with the proof by Gauss, two hundred years later, that the main field could only be of internal origin, it amounts to one of the most far-reaching discoveries in Geophysics. It is a demonstration that within the Earth there is a relative motion of its parts. Although slow by comparison with human activity it is a million times faster than large scale geological changes. We have probably not yet explored all the consequences of this discovery.

The longest records of secular variation are from London and Paris, which are sufficiently close together to show substantially the same long term changes. The variation at London is represented two-dimensionally in Figure 42 to show inclination and declination simultaneously. Similar plots from other observatories, although based on shorter records, generally show the same sort of variation. Figure 42 is suggestive of a cyclic variation with a period of about 500 years, but archaeological evidence (page 164) does not allow us to conclude that there is really a regular cycle of secular variation.

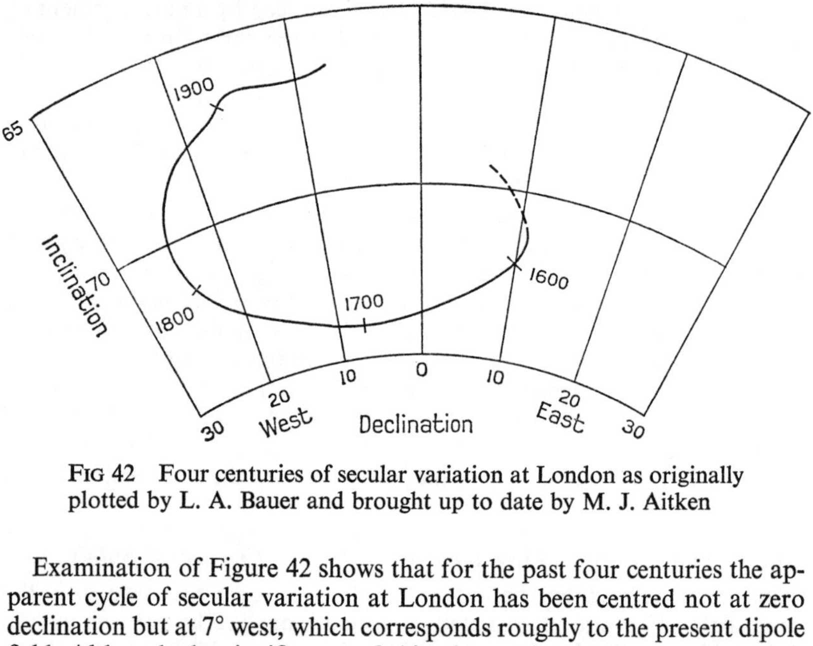

FIG 42 Four centuries of secular variation at London as originally plotted by L. A. Bauer and brought up to date by M. J. Aitken

Examination of Figure 42 shows that for the past four centuries the apparent cycle of secular variation at London has been centred not at zero declination but at 7° west, which corresponds roughly to the present dipole field. Although the significance of this observation is now questioned, it has played a very important role in the development of theories of the origin of the main field. It indicates that the dipole component of the field has remained nearly constant in orientation, implying that the secular variation is due primarily to variability in the non-dipole field. Analyses of the whole field at different intervals confirm that, although the dipole field is not constant in direction and is decreasing in magnitude at the present time, the variability of the non-dipole field is greater and that it is the non-dipole field which dominates the pattern of secular variation. Archaeological and palaeomagnetic evidence (Chapter 4.2) indicates that this situation may be characteristic only of the few centuries of direct geomagnetic observations, but it is nevertheless a convenient starting point for a discussion of the secular variation and the origin of the main field.

An analysis of the secular variation by E. C. Bullard* and others showed that the whole of the non-dipole field, as plotted in Figure 41, drifts slowly westwards, and that a large part of the secular variation as seen at one point is due to the passing 'highs' and 'lows' of the westward drifting non-dipole field. In the interval 1907 to 1945, they found that the westward drift had occurred at an average rate of 0·18 degrees of longitude per year, being the same at all latitudes. At this rate the whole pattern of the non-dipole field would complete one revolution of the Earth in 2,000 years, and since the characteristic pattern of the non-dipole field around 50° latitude in Figure 41 is a succession of two highs and two lows, we would expect the London secular variation to show a cyclic tendency, with two cycles per 2,000 years, i.e., 1,000 years per cycle, instead of 500 years per cycle as apparent in the observed curve (Figure 42). The disagreement shows that the westward drift is not sufficient to explain the secular variation. In addition to the general westerly drift the pattern of the non-dipole field is undergoing continual change, so that we could not expect any one feature to survive even one complete circuit. The magnetic field has been compared with the atmospheric circulation in which high or low pressure areas develop and drift generally *eastwards*, but change their forms, continuously producing new features to replace those which disappear. Moreover, palaeomagnetic evidence makes it clear that the main dipole field is not as constant in direction as the past few decades of records make it appear to be.

Historically the westward drift has been a vital factor in the develop ment of a theory of the geomagnetic dynamo, although its fundamental significance is now doubted. It was interpreted as evidence that the outer part of the core, in which the non-dipole field originates, is rotating more slowly than the mantle or solid part of the earth, and that since the mantle was coupled electromagnetically to the core as a whole, the outer and inner parts of the core were rotating at different speeds. Differential rotation within the core almost certainly occurs but it now appears unlikely that this is the cause of the westward drift, which is more likely to be a statistical feature of the present field configuration and may well change to an eastward drift over a few thousand years.

Generation of the main field

Our detailed picture of the physical state of the interior of the Earth is the principal result of the study of seismology. The outer 2,900 km of the Earth, the mantle, is composed of solid material with stony composition, which transmits shear waves as well as compressional waves. The inner sphere,

* E. C. Bullard, C. Freedman, H. Gellman and J. Nixon, 'The westward drift of the earth's magnetic field', *Proc. Roy. Soc.* (London) A, Vol. 243, p. 67 (1950).

the core, shown to scale in Figure 40, does not transmit shear waves at all and must therefore be liquid. Its density is much greater than that of the mantle and there is virtually no alternative to a composition of liquid iron with minor proportions of dissolved elements, particularly nickel. The theory of the generation of the main field, developed by W. M. Elsasser in U.S.A. and E. C. Bullard in England and subsequently elaborated by many others, depends upon the fact that the core is a rotating, liquid electrical conductor.

It has usually been assumed that the internal motions of the core which are responsible for its dynamo action are driven by thermal convection due to heat generated in the core. Originally this was attributed to the radioactive decay of traces of uranium, thorium and potassium. Although we have no means of sampling the Earth's core to determine whether this is possible, it is most likely that the core is similar in composition to the iron meteorites (see page 173) in which case the concentration of radioactive elements is far too small. An attempt was made to save the convective dynamo by postulating that the whole Earth is cooling down at a rate such that over geological time the inner, solid core had progressively grown from nothing and that the latent heat of solidification so liberated had driven the convection of the outer, liquid core. Recent quantitative consideration of this postulate, shows however, that it too is inadequate.* The driving mechanism which must now be favoured is the precession of the Earth, as has been pointed out by Malkus†.

Precession of a spinning body occurs when it experiences a torque normal to the axis of its rotation. In the familiar case of a gyroscope, the torque results from the fact that gravity, effectively acting through the centre of gravity is not in line with the vertical force acting at the point of support. Precession of the Earth is a consequence of the gravitational gradients of the Moon and Sun acting on the Earth's equatorial bulge. The plane of the Moon's orbit about the Earth is close to the plane of the Earth's orbit about the Sun (ecliptic), but the plane of the Earth's equator is misaligned by $23\frac{1}{2}°$ from the ecliptic. It is convenient to refer to the $23\frac{1}{2}°$ difference between the axis of rotation and the normal to the ecliptic. The precession is a slow rotation of the polar axis about the normal to the ecliptic, as in Figure 43. It is important to note that the amplitude of the Earth's precession is not small, although it is slow; a complete cycle takes 26 000 years. The relevance of the precession to the geomagnetic field is that its rate is determined by the magnitude of the equatorial bulge; but the bulge of the core is not the same as that of the earth as a whole, so that the core tends to precess at a different rate. Thus if the core and mantle (solid outer part of the Earth) were mechanically uncoupled their axes of rotation would precess independently about the $23\frac{1}{2}°$ cone and at their extreme deviation

* See F. D. Stacey, *Physics of the Earth* (Wiley, New York; 1969).

† W. V. R. Malkus, Precession of the Earth as the Cause of Geomagnetism. *Science,* Vol. 160, p. 1048 (1968).

would differ by 47°, in which case there would be a velocity difference of 175 metres/sec between the core and mantle at their boundary. In fact the core and mantle are coupled electromagnetically by virtue of the electrical conductivity of the mantle, as discussed below, and the core is forced to follow the mantle precession, but it does so with a lag which increases inwards. The effect is rather like stirring the core.

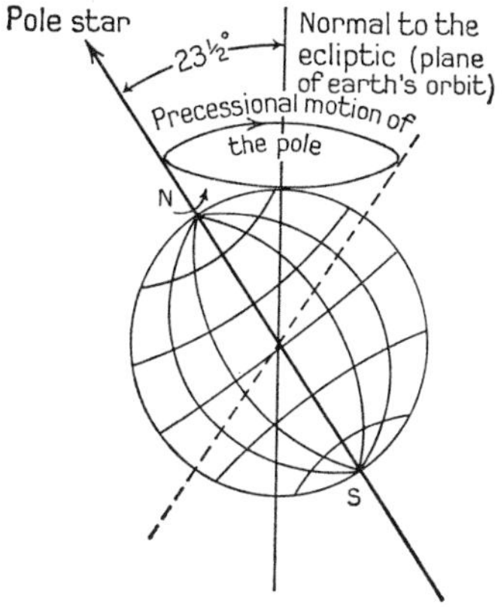

FIG 43 Precession of the Earth; the pole follows a cone of semi-angle $23\frac{1}{2}°$ with a period of 26 000 years

Given a satisfactory cause for large scale motions within the core, the principles of hydromagnetic dynamo theory, developed to explain field in terms of convective motion, can still be used. The essential point is that the liquid iron, of which the core is believed to be composed, is a good electrical conductor, and movement of a magnetic field through an electrical conductor generates circulating electric currents which oppose the motion of the field; it would take several thousand years to move a magnetic field through a conductor the size of the core, so that the magnetic field lines are effectively frozen in to the core and become distorted as it moves. Figure 44

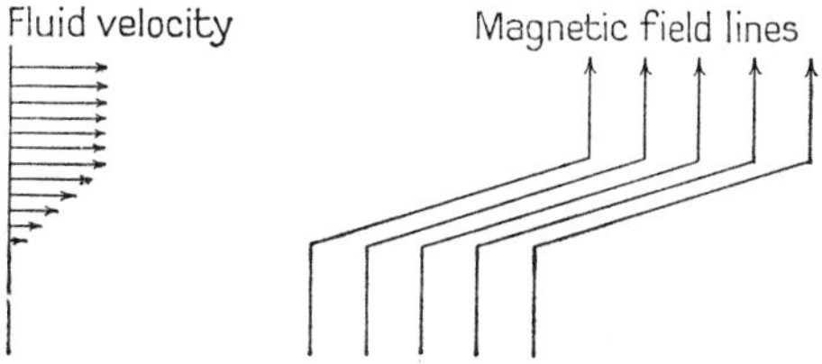

FIG 44 Illustration of the intensification of a magnetic field by a shear velocity in a fluid conductor. Field strength is represented by the closeness of the field times. This process converts kinetic energy of the fluid motion into magnetic energy of the field

indicates the intensification of a magnetic field by a velocity shear in a fluid conductor to which the field lines are frozen.

If we start with the assumption that there is a dipole field within the core, then circumferential velocity shears arising from the precessional torques twist the field lines, producing, in effect, an additional field of toroidal form, that is, having field lines forming closed loops within the core. The existence of a toroidal field, of which we can have no direct evidence because it is necessarily confined to the core in which it is produced, is an essential feature of all theories of the geomagnetic dynamo. Radial motions occurring in the core, probably in turbulent eddies apparent as features of the non-dipole field, deform the toroidal fields, producing fields of dipole form, which can, with an appropriate pattern of motion, reinforce the original dipole field. A state of zero field is not stable; any disturbance triggers dynamo action.

We can also analyse the geomagnetic field in terms of the electric currents in the core which are required to maintain the field. The ohmic dissipation in the core and the necessity for a very large power source to maintain the field then become obvious. The energy for the precessional dynamo is drawn from the Earth's rotation and probably accounts for as much as one third of the slowing of the rotation (the dominant effect being the marine tides).

Consequences of electrical conduction in the mantle

The field must be regarded as locked to the core which is a much better electrical conductor than the mantle. Any motion of the mantle with respect to the core therefore involves moving the mantle through the field and induces electric currents in the mantle which react with the field to oppose the relative motion and, after a time, stop it altogether unless it is maintained by a mechanism which supplies the energy dissipated by the currents. But since there is a relative motion of the dipole and non-dipole fields and the mantle is coupled to both, there is a continuous dissipation and the secular variation, in particular the westward drift, is a result of a dynamic balance in the coupling of the mantle to the various features of the field. It is therefore apparent that the conductivity of the mantle as well as of the core is important in understanding the details of geomagnetic observations.

Since the mantle is presumed to be of stony composition and rocks are reasonable insulators under normal laboratory conditions, appeal is made to Solid State Physics for an explanation of the onset electrical conduction at the extreme temperatures and pressures of the lower mantle. There is no possibility of reproducing in the laboratory the actual conditions prevailing in the lower mantle (two million atmospheres and 3,500°C) but the extrapolation of measurements made at more moderate temperatures and pressures yields estimates of electrical conductivity in accord with geophysical requirements.

The maximum conductivity at the bottom of the mantle is in the range 1 to 10 ohm^{-1} cm^{-1}, which is typical of materials known as semi-conductors, being about 10^4 times smaller than the conductivities of metals. In a semi-conductor at low temperatures the electrons are bound to the atoms and are not free to migrate from one atom to the next, but as the temperature is raised they may be excited by thermal agitation out of the bound state into a conduction band, which is a more energetic condition, allowing them move more less freely through the solid. With increasing temperature the number of electrons excited into the conduction band rises very sharply and the electrical conductivity increases correspondingly. This is the principal reason for the general rise in conductivity with depth in the mantle. Pressure also plays a part, by reducing the energy gap which the electrons have to jump between the bound and conducting states in order to contribute to the conductivity. At very high temperatures ionic conduction (as distinct from electron conduction) has also to be taken into account, but the movement of ions is inhibited by pressure so that it is doubtful whether ionic conduction becomes significant in the mantle. Figure 45 shows two

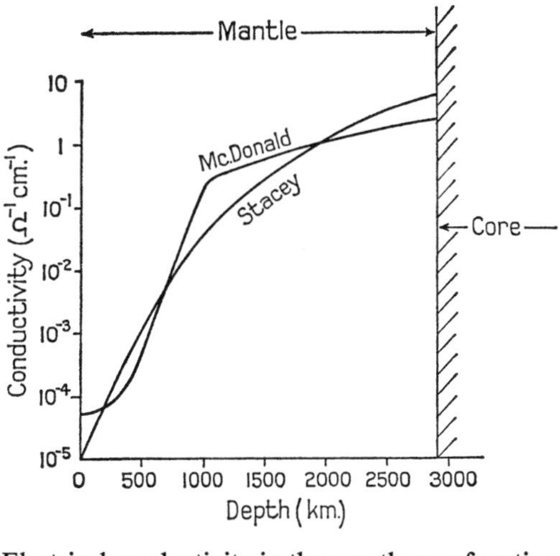

FIG 45 Electrical conductivity in the mantle as a function of depth according to McDonald* and Stacey†

estimates of conductivity as a function of depth within the mantle, both of which are consistent with Solid State and Geophysical considerations.

A rough upper limit to the conductivity of the lower mantle is set by the fact that the secular variation can be observed at the surface at all. When an

* K. L. McDonald. Penetration of the geomagnetic secular field through a mantle with variable conductivity. *Journal of Geophysical Research*, Vol. 62, p. 117 (1957).

† F. D. Stacey. *Physics of the Earth* (Wiley, New York; 1969).

electrical conductor is exposed to a changing magnetic field the eddy currents which are induced in the conductor oppose the penetration of the field changes. In the geometrically simple case of a conductor which is infinite in extent below a plane surface where a magnetic field varying sinusoidally with time is imposed upon it, the field is attenuated exponentially with depth of penetration. If the field at the surface has amplitude H_0 and frequency f(Hz) then at a penetration depth z (cm) the amplitude is $H_0 e^{-2\pi z\sqrt{f\sigma}}$ for a uniform conductivity σ in electromagnetic units. The attenuation is a function of frequency, being greater for higher frequencies. Thus if we consider the penetration of a field which fluctuates irregularly, as it seems probable that the Earth's field does at the surface of the core, we find that the more rapid fluctuations are attenuated to a greater extent. The secular variation at the surface of the Earth therefore gives us a smoothed-out picture of a fluctuating field of the core, not merely because we are looking at it from a distance, but because we are looking through the electrically conducting mantle. It is apparent that secular changes occurring over a few years do just penetrate the mantle and therefore that if the mantle were a uniform conductor its conductivity could not exceed about 1 $\text{ohm}^{-1}\ \text{cm}^{-1}$. Allowing for the increase of conductivity with depth, the value at the bottom of the mantle is likely to be about 5 $\text{ohm}^{-1}\ \text{cm}^{-1}$.

Having obtained a reasonable estimate of the conductivity of the lower mantle we can consider further its electro-mechanical coupling to the core. The most direct evidence of mechanical coupling comes from measurements of fluctuations in the length of the day. Changes in the rate of rotation of the Earth (that is, the solid part which can be observed) amount to a few parts in 10^8, that is one second in the length of a year. Astronomical observations over the past three centuries indicate that the changes are erratic in nature and may occur slowly over many years or more rapidly in a few years. Since there is no external agency capable of causing the total angular momentum of the Earth to fluctuate in this way, the changes must be ascribed to variations in moment of inertia or to the internal redistribution of angular momentum between the core, mantle and atmosphere. Variations in the moment of inertia of the Earth must be allowed as a possible cause for slow changes in the rotation (but would in this case be difficult to distinguish from damping due to lunar tides), but cannot be adequate for the more rapid changes. They would require, for instance, the melting or accumulation of Antarctic and Greenland ice caps sufficient to produce a world-wide change in sea level of six inches in a few years, which certainly does not occur.

We must also discount the atmosphere which has too small a mass to influence the rotation to this extent. Its mass is 10^{-6} of that of the whole Earth and its moment of inertia a little more than 2×10^{-6} of that of the whole earth. Thus to cause a change in rotation of the mantle by a few parts in 10^8 the angular momentum of the atmosphere would have to change by at least 1 per cent and this corresponds to an average world-wide wind (easterly or westerly) of about 6 knots. While changes in atmospheric circulation are occurring continuously there is no evidence of any average changes as gross as this.

The only reasonable explanation for changes in the Earth's rotation therefore requires the exchange of angular momentum between the core and the mantle. Since they are magnetically coupled, the coupling must fluctuate; this would occur by virtue of fluctuations in the relative strengths of different features of the field associated with irregularities in core motion.

We can make a reasonable estimate of the tightness of the coupling from the length-of-day observations which indicate that if the core-mantle exchange is disturbed it returns to equilibrium in about five years. Calculations based on dipole and non-dipole fields of about four oersteds at the core-mantle boundary and associated toroidal fields arising from the differential rotation of the core and mantle show that a coupling time-constant of five years is possible with a lower mantle conductivity of about 2 ohm^{-1} cm^{-1}, as indicated by other estimates.

Evidence that the mantle becomes electrically conducting at depth has also been obtained from the much more rapid geomagnetic fluctuations which are due to the effects of solar activity on the outer reaches of the geomagnetic field (the magnetosphere) and to the rotation of the Earth with respect to the Sun. The causes of these fluctuations are discussed in Sections 4.3 and 4.4, but we may note here that, assuming the current flow through the surface of the Earth to be negligible a harmonic analysis of the disturbance field over the surface can be used to separate those fractions which are of external and internal origin. An appreciable internal part is always found and its correlation with the external part compels us to explain it in terms of deep-seated electrical currents which are themselves induced by the fluctuations. Although this does not allow us an explicit determination of conductivity as a function of depth within the Earth, the distributions of conductivity which are physically plausible as well as being consistent with the observations are quite limited. Below 600 to 700 km the conductivity must increase very sharply and reach a value not less than 10^{-2} ohm^{-1} cm^{-1} at 1,000 km. The disturbances are too rapid to penetrate significantly beyond that, so that no information about the lower mantle is obtainable in this way. However, the possibilities of drawing a smooth curve through acceptable values for both the upper and lower mantles (as in Figure 45) are quite restricted.

4.2 Palaeomagnetism — The Pre-history of the Earth's Magnetic Field

The study of rock magnetism can be traced back to the late 19th century, but its current significance has been apparent only since about 1950. It was then becoming widely recognized that many rocks have preserved a fossil record of the geomagnetic field extending back not merely for thousands of years, but for hundreds and perhaps thousands of millions of years. A new subject, palaeomagnetism or 'old magnetism,' emerged. After 15 years of careful collection and collation of data,* it has caused a revolution in geophysical thinking which is remarkable both for its suddenness and its completeness.

Origins of rock magnetizations

An essential feature of the art of making stable permanent magnets is the production of a very fine grained magnetic material. Nature has produced such material on a grand scale. Almost all rocks contain small grains of minerals such as magnetite and hematite which cause the rocks to be appreciably magnetic. It is only very rarely that these minerals occur in sufficient concentration to produce a rock such as lodestone, which is itself a usable magnet, but this is because the concentrations of magnetic minerals are normally low (0·1 to 10 per cent). The important feature of rock magnetizations is that, although normally slight in intensity, they are magnetically hard, that is, remanence is not easily destroyed. In particular the primary remanent magnetization which is induced in a rock at the time of its formation is generally preserved indefinitely. Commonly additional, secondary magnetizations are induced subsequently, but can be recognized as secondary because they are less stable. The problem of palaeomagnetism is to distinguish the primary magnetization and from its direction (and perhaps strength) to deduce the direction (and strength) of the ancient geomagnetic field.

Of particular relevance to palaeomagnetism is the process of thermoremanent magnetization, by which igneous rocks acquire their primary remanences. We consider a body of igneous rock which is uniform and

* E. Irving, *Paleomagnetism* (Wiley, New York; 1964).

isotropic and cools slowly from its solidification temperature to normal atmospheric temperature. Its magnetism first appears at a transition temperature, known as the Curie point (generally between 400°C and 580°C, depending upon the composition of the dominant magnetic mineral) and immediately assumes the direction of the magnetic field in which the rock is situated. However, if the field is removed while the rock is still very close to its Curie point, the magnetization disappears. This is because it is only an induced magnetization, not a remanence and the magnetic domains are agitated thermally and lose their net alignment. If the rock is cooled in the field its magnetization becomes stable at a *blocking temperature* a few tens of degrees below the Curie point. Thereafter further cooling increases the stability of the magnetization which has been induced and any subsequent changes in the field have no effect. The remanence is 'frozen in' by the cooling and, since it was induced at a high temperature, it is known as *thermoremanence*. The stability of thermoremanence in a rock in a property of the rock and is independent of the intensity of the field in which is it induced Thermoremanence induced in the Earth's field (about 0·5 oersted) can only be destroyed by a much stronger alternating field, usually several hundred oersteds, whereas a magnetization induced at room temperature is destroyed by an alternating field of the same magnitude as the inducing field.

Sedimentary rocks are also used for palaeomagnetism. They commonly contain grains of magnetite which have carried with them thermoremanent magnetic moments, induced during their previous, igneous history, which align themselves with the geomagnetic field during deposition and thereby impart a remanent magnetization to the sediment. This is known as *detrital magnetization*. Frequently, however, sediments undergo chemical changes during consolidation and the initial detrital remanence is replaced by a new magnetization, known as *chemical remanence*. The original magnetite (Fe_3O_4) grains are lost in this process and the iron is converted to hematite (Fe_2O_3), the fully oxidized state, usually via a hydrated oxide. Since the hematite is formed at temperatures well below its Curie point the magnetic domains into which it is divided assume the direction of the field and the stability of the resulting remanence is similar to that of thermoremanence.

The primary remanences of rocks are commonly sufficiently stable to have survived without serious decay from the time of their formation to the present day. Of course, if they have been altered by re-heating, stressing or chemical metamorphism then this will not be true, so that metamorphic rocks are avoided in palaeomagnetism, but even with igenous and sedimentary rocks secondary magnetization may be induced by the very prolonged exposure of the rocks to the geomagnetic field. This is possible because the alignment of magnetic domains, which is responsible for the natural remanence of a rock, is normally very slight, of the order 0·1 per cent of magnetic saturation or complete alignment. Domains of lower than average magnetic stability, which may or may not contribute to the primary remenance, gradually align themselves with the geomagnetic field at a time much later than the formation of the rock, when the field may have a direction quite different from that of the primary remanence. This process is known as viscous magnetization. Any *viscous remanence* must be removed before the direction of a primary remanence can be measured.

The methods of palaeomagnetism

Since the magnetism of rocks is slight, sensitive instruments are needed to measure it. They are of two types, the astatic and the spinner or rock-generator magnetometers illustrated diagrammatically in Figures 46 and 47.

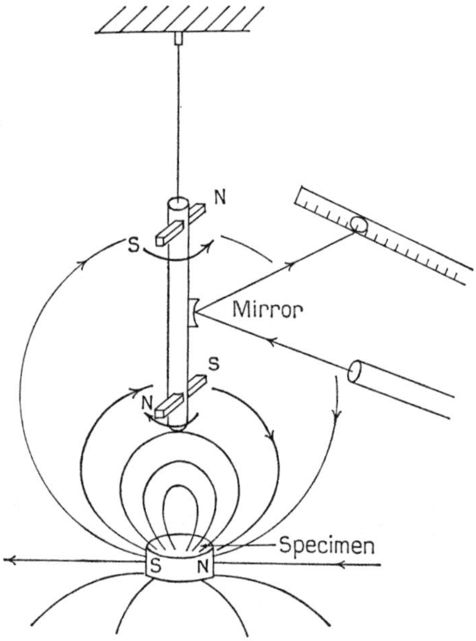

FIG 46 Principle of the Astatic Magnetometer

A simple astatic magnetometer consists of two small permanent magnets of equal strength, which are rigidly held apart by a light rod. Their magnetic axes are opposite and perpendicular to the rod, which is suspended by a fine torsional fibre or strip. With this arrangement any uniform field produces equal and opposite torques on the two magnets, causing no rotation. However, a magnetized specimen held underneath the magnet system at a distance comparable with the magnet separation has a much greater effect upon the lower magnet. The deflection which it produces is a measure of the magnetic moment of the specimen, which has to be held in a series of positions and orientations in order that the magnitude and direction of the remanence can be calculated from the corresponding deflections. Disturbance by induced magnetism in the Earth's field is avoided by placing the whole apparatus at the centre of a system of large coils in which the currents are so adjusted that they cancel the Earth's field in the central volume. This volume is referred to as a zero field space or field-free space.

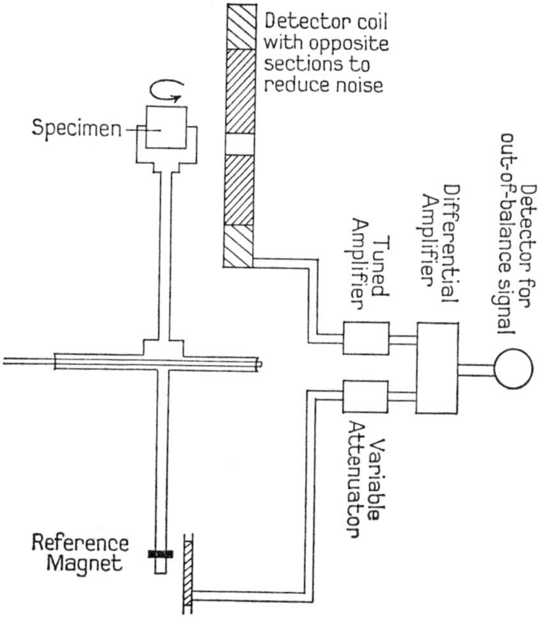

FIG 47 Principle of the Spinner Magnetometer

One form of spinner magnetometer is shown in Figure 47. The rock specimen is rotated close to the centre of a large coil in which it induces a voltage at the frequency of rotation, with an amplitude proportional to the component of the magnetic moment which is perpendicular to the axis of rotation. The direction of magnetization is determined by the phase of the voltage with respect to a reference signal, which may be generated by a small magnet rotating on the same shaft or by painting the specimen holder so that it modulates a light beam to a photocell.

Both types of instrument have been developed to measure weakly magnetic sediments, with moments as small as 10^{-6} electromagnetic units per cubic centimetre, as well as the more strongly magnetic igneous rocks.

Before the measured natural remanence of a rock can be interpreted as a primary remanence, indicative of the ancient geomagnetic field, viscous or secondary magnetizations are removed by a process of partial demagnetization. Viscous remanence is much more readily removed by an alternating field than is thermoremanence so that, at least for most rocks, it is possible to find an alternating field intensity which will destroy the secondary remanence but leave most of the primary remanence. Specimens are subjected to an alternating field which is slowly decreased to zero from the selected intensity. Demagnetization is carried out in a zero-field space and the specimens are rotated at different frequencies about two or three mutually perpendicular axes simultaneously, so that the specimens are exposed to the demagnetizing field in all orientations and no steady (magnetizing) field is superimposed upon the alternating (demagnetizing) field.

Rocks can also be demagnetized thermally by heating and cooling them in field-free space. The thermal method is generally less convenient than the alternating field method, but can be more useful when applied to rocks which may have complex thermal histories. By observing the progressive demagnetization at a series of temperatures, the history can be re-traced in reverse.

Extension of the record of secular variation—archaeological evidence

Direct observations of the secular variation extend back only 400 years and this is too small a time span to allow us to make more than a guess at the longer term behaviour of the Earth's field; the obvious guess is wrong. The record has been extended by applying palaeomagnetic methods to dated archaeological material, historically recorded lava flows and, with rather less precision, to dated sediments. Most clays contain several per cent of iron oxides so that vases and bricks acquire thermoremanence as they cool after firing. Early pottery appears always to have been baked upright and it therefore records the inclination of the geomagnetic field at the time, but not the declination because its orientation in the horizontal plane was arbitrary. Rather more useful are the brick-built pottery kilns, for which approximate dates of last firing are known from radio-carbon measurements in many cases. Since many have remained undisturbed, they record both inclination and declination of the field in which they last cooled.

Although the archaeomagnetic record is fragmentary, it is sufficient to discount the regular cyclic secular variation which is implied by the past few centuried of direct observation. The secular variation curve deduced by Dr. M. J. Aitken and co-workers at Oxford from measurements on brick samples from dated pottery kilns in England, is reproduced in Figure 48, from which it is evident that secular variation is quite irregular and even the sense of the rotation of the field vector changes.

Past variations in the strength of the Earth's field have received much less attention than variations in direction. This is partly because intensities of magnetization are much more affected by viscous changes than are directions and the uncertainty in measurement is therefore greater. Nevertheless archaeomagnetic measurements, particularly those in France and Japan indicate that the field has decreased in intensity by a factor of about 1·5

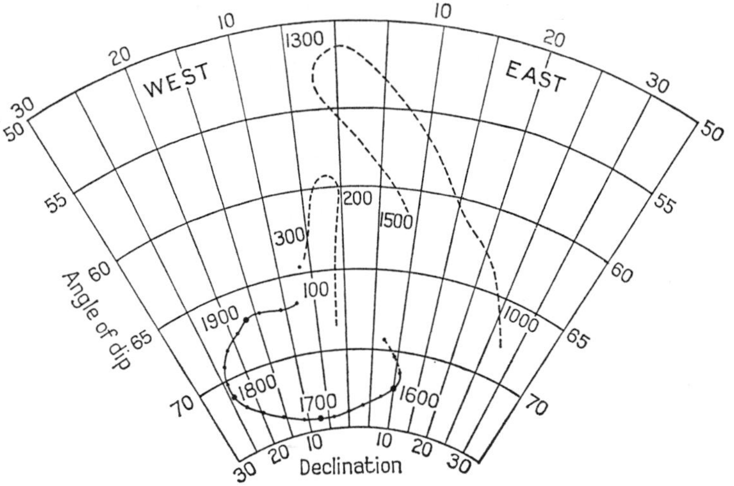

FIG 48 Secular Variation in England, deduced from the remanent magnetism of bricks from dated pottery kilns and compared with the direct observations since 1580. Figure from data by M. J. Aitken and others

over the past 2,000 years. The magnitude of this change is sufficient to preclude its explanation as a local variation of the non-dipole field. It is also consistent with direct measurements over the past century which show that in this time the *dipole* field has decreased in intensity by several per cent. Thus it is not only the features of the non-dipole field which wax and wane. The whole dipole field evidently does so too, probably quite irregularly. This observation carries the interesting implication that the geomagnetic dynamo has a certain instability, its extreme fluctuations being complete reversals in the polarity of the field (page 167).

The geocentric axial dipole hypothesis

The principles of symmetry are fundamental laws of Physics which have never been seriously questioned although they are rarely enunciated because in most cases their applicability is sufficiently obvious to make any detailed consideration trivial. However, one of them has an important application to the geomagnetic field. It states that no effect can have a lower symmetry than the combination of its causes. Since a dipole field has axial symmetry (it also has a polarity but discussion of this point is deferred to the following section), we can look for axial symmetry in the causes of the Earth's field. The only property of the Earth in which such symmetry is evident is its rotation which, as discussed on page 154, plays a vital role in generating the field. Therefore, unless some important factor has been omitted from the theory, the axis of the geomagnetic field must on average coincide with the geographic or rotational axis. That it does not do so at the present time does not constitute a paradox as long as we regard the present departure of the field from the geographic axis as transient.

Even without the changing form of the non-dipole field, the rate of its westward drift ensures that when observations of the geomagnetic field are averaged over a few thousand years the non-dipole field is averaged out, leaving only the main, dipole field. In palaeomagnetism, measurements are normally made on numerous samples which embrace a time interval at least this long (i.e., sequences of lava flows or a thick layer of sediment), and therefore the average field direction deduced from such a series of measurements is that due to the dipole field over the interval when the rocks were formed. The striking result of palaeomagnetic work on rocks which are geologically young (late Pleistocene and Recent in geological parlance) is that the average apparent geomagnetic poles deduced by several independent workers cluster convincingly about the geographic pole, leaving the present magnetic pole very much on one edge of the distribution (Figure 49). Careful statistical treatment of the results confirms the coincidence with the geographic poles of the average magnetic poles over a time span of several thousand years.

This conclusion is of the greatest importance in palaeomagnetism, it leads directly to the hypothesis that, except perhaps for brief periods when it was undergoing a reversal in polarity (see page 167) the geomagnetic field averaged over a few thousand years, at any period of geological time, was a simple, dipole field, centred at the centre of the Earth with its axis coincident with the axis of rotation. This is the Geocentric Axial Dipole Hypothesis.

Although the treatment of earlier geological periods can never be as reliable as the most recent, measurements on rocks formed during limited periods (a few million years) at widely spaced sites in individual continents indicate that the geomagnetic field was dipolar also in the remote past. It is the conclusion that measured palaeomagnetic poles refer not merely to the average geomagnetic pole at earlier geological periods but to the pole of rotation at those periods which has put the hypotheses of polar wander and continental drift on to a quantitative basis.

The scatter of pole positions calculated from the magnetizations of individual rock samples from a single geological formation (sequence of sediments, series of lava flows or even a single large flow which cooled slowly) is normally much greater than the errors of measurement and is attributed to secular variation about the mean or dipole field. The generally similar palaeomagnetic scatter at all geological periods indicates that the secular variation has always had a similar magnitude, but this conclusion must still be treated with caution because the possibility that there are other significant causes of palaeomagnetic scatter has not received detailed attention.

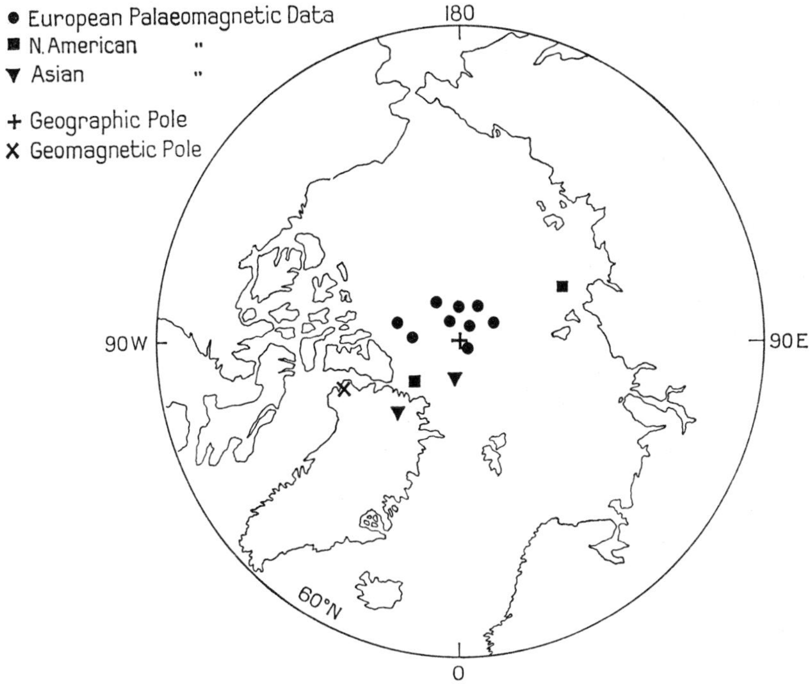

FIG 49 Late Pleistocene and Recent Virtual Geomagnetic Poles. The results of several workers represented on a single figure

Reversals of the Earth's field

One of the early discoveries of palaeomagnetism was the fact that many rocks are magnetized in a direction approximately opposite to that of the present field. The simplest explanation is that they were formed at a time when the polarity of the field was reversed, but this explanation was at first treated with caution as there had previously been no reason to suspect that the earth's field was ever reversed. The subsequent discovery, notably at Mt. Haruna in Japan, of rocks whose magnetic remanence was self-reversing (that is, they were found to acquire thermoremanence opposite to the polarity of the inducing field) caused considerable doubt about the significance of the reversals. Careful study of the self-reversal mechanism and an overwhelming accumulation of palaeomagnetic evidence now leave little doubt that self-reversals are rare and that reversals of the Earth's magnetic field have occurred many times. The discovery by Drs. H. W. and H. D. Babcock at the Mt. Wilson and Palomar Observatories in California, that the sun's magnetic field reverses during the 11-year solar cycle, reinforces the explanation of terrestrial reversals as alternations of a hydromagnetic dynamo.

Strong circumstantial evidence favouring reversals of the field rather than self-reversing rocks was provided first by sequences of lava flows in Iceland and Oregon. In both places numerous distinct flows are superimposed in what is clearly chronological order. They show alternating groups with normal and reversed magnetizations. There is no obvious physical or chemical distinction between the normal and reversed rocks which would be suggestive or a self-reversing process occurring in one but not the other. The numbers of normal and reversed rocks are approximately the same, which clearly indicates that the Earth's field may have either polarity with equal probability. A self-reversing mechanism which happened to operate in just 50 per cent of the rocks would be a very remarkable chance occurrence.

Even more convincing evidence is provided by Dr. R. L. Wilson's analysis of the data on all of the *baked contact rocks* for which magnetic polarities have been reported. A baked contact occurs when hot, igneous rock makes contact with an older cold rock. The surrounding rock which is remote from the new igneous mass retains the remanence which dates from its own formation but the rock immediately adjacent to the fresh lava is baked, that is heated to a temperature above its Curie point, so that its remanence disappears. It then acquires a new remanence as it cools with the new igneous rock. Both normal and reversed polarities were found, but with three doubtful exceptions out of 52 reported baked contacts, the magnetic polarities of the baked rocks coincided with those of the igneous rocks which heated them independently of the polarities of the unbaked country rocks. This is what one expects if self-reversals are rare and reversed magnetization is due to field reversals. If the reversed polarities had been due to self-reversals then we would not expect that every time a self-reversal occurred in an igneous rock, it occurred also in the baked contact rock, which is usually quite different physically and chemically. We would expect instead that the baked and unbaked rocks would agree in polarity, which they do in only 50 per cent of the cases, again reflecting the fact that the geomagnetic field may have either polarity with equal probability.

The alternating polarity of the field means that on the long time scale its symmetry is axial but not polar. There is therefore no connection between the sense of the Earth's rotation and the polarity of the field. According to the hydromagnetic theory, a field with either polarity is self-maintaining. However, the cause of reversals cannot be made the subject of a detailed theory. Reversals of the solar magnetic field are evidently more or less regular and are best explained as an intrinsic property of the hydromagnetic dynamo and models have been devised which can simulate it. Reversals of the Earth's field are probably irregular which makes their explanation more problematical, but an instability of the dynamo system or perhaps a disturbance in the lunar or terrestrial orbits must be assumed.

It appears now that self-reversals are virtually if not entirely confined to rocks whose magnetic minerals are solid solutions of hematite and ilmenite (a similar mineral in which some of the iron atoms are replaced by titanium). After the mineral grains have acquired thermoremanence in the same sense as the field a process of ionic ordering occurs, in which the magnetically active iron atoms and inactive titanium atoms change places to occupy favoured positions in the crystal structure instead of the random positions to which they were driven thermally at higher temperatures. The orientations of the magnetic moments of the iron atoms in their new sites are determined by the interactions with neighbouring atoms being in many cases opposite to their initial orientation, so that if the process goes far enough, the total thermoremanance may be reversed.

A consequence of geomagnetic reversals with very far-reaching scientific implications has emerged from magnetic surveys of the oceans. Characteristic features of the ocean floors are the ridges, of which the mid-Atlantic ridge is the best known and most striking example. It neatly bisects the Atlantic ocean from the Arctic to the Southern Ocean; it is a seat of minor seismicity over most of its length and is obviously an important, active feature of the Earth's crust. Magnetic surveys have revealed linear magnetic anomalies parallel to the ridges in all oceans. These are now recognized as direct evidence of sea-floor spreading from the ocean ridges, as in Figure 50. Fresh mantle material is rising to the ocean floor along the ridges and

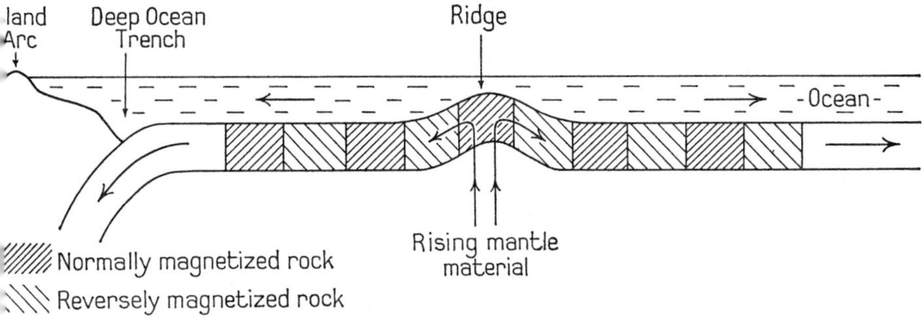

FIG 50 Diagrammatic representation of the sea floor spreading hypothesis. As fresh mantle material rises to the ocean floor along a ridge it cools and becomes magnetized in the direction of the prevailing geomagnetic field. Continued outward spreading of fresh ocean floor material during many periods of alternating geomagnetic polarity has produced a series of magnetic 'stripes' parallel to the ridge and these are apparent in magnetic surveys, which show the symmetry of the magnetic pattern about the ridges. Dating of the principal polarity epochs has allowed the rates of sea floor spreading along several ridges to be estimated as a few centimetres per year

spreading outwards, the magnetic 'stripes' indicating the polarity of the geomagnetic field as the material emerged. The process is driven by 'convection' in the mantle and is a vital part of the global tectonic pattern.*

* B. Isacks, J. Oliver and L. R. Sykes, Seismology and the new global tectonics. *Journal of Geophysical Research*, Vol. 73, p. 5855 (1968).

Polar wander and continental drift

We are here concerned only with the average magnetic pole over millions of years so that the geocentric axial dipole hypothesis is assumed. The word 'pole' therefore means both the magnetic pole and the geographic pole, the two having become coincident on this time scale. We are concerned with changes which occur slowly, over hundreds of millions of years, so that even reversals appear frequent and the polarity is disregarded; we are simply considering movements of the axis of rotation with respect to the surface features.

Poles calculated from measurements on rocks younger than about 20 million years do not differ from the present geographic pole by distances greater than the experimental uncertainties. Beyond that the differences certainly exceed the limits of experimental error and become progressively greater the further back the record is taken. Considering first the results of measurements on rocks from a single continent, we can plot on a globe, or on a stereographic projection of it, the position of the pole corresponding to each geological period. A curve through these poles is a polar wander curve (for that particular continent), a dated path which the pole has followed. Its rate of movement appears to average about one third of a degree per million years or three centimetres per year, but it may well be quite irregular. It now seems likely that polar movement occurs during geologically short periods of activity separated by longer quiescent intervals.

As soon as we consider more than one continent, simple polar wander is no longer sufficient to explain the observations. The first two polar wander curves to be compared were those of Western Europe and North America. The striking fact is that although they have the same general shape they are not coincident. They converge to the present pole from a separation of about 3,000 miles 300 million years ago. By the mid-1950's it was generally recognized that if North America were joined to Europe 300 million years ago but separated and drifted gradually westward to its present position leaving the Atlantic Ocean in the gap, the two curves would be reconciled. This is not a unique solution to the geometrical problem but it emphasizes that the palaeomagnetic data can only be explained by relative movements of continental blocks by thousands of miles. This is continental drift, a phenomenon postulated many years earlier for quite different reasons but almost universally discounted. In fact the resurrection of the theory of continental drift was so disturbing to established geophysics that in one prominent geophysical laboratory the study of palaeomagnetism was deliberately suppressed. Polar wander curves for three continents are shown in Figure 51. The movement of Australia relative to the northern continents is even more striking than the relative movement of Europe and North America, and leaves no doubt about the significance of the discrepancies in polar wander curves of different continents.

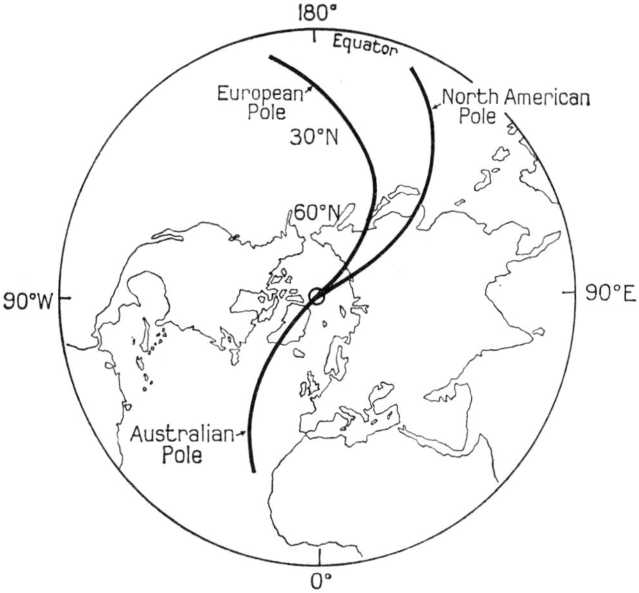

FIG 51 Polar wander curves for Europe, North America and Australia from the Silurian to the present time

Many years before the intensive study of palaeomagnetism began, geological evidence of polar wander from ancient climates (for example coal seams in Antarctica) was already strong. Even the large angular migration of the pole which must be considered (at least 90° in 300 million years) caused no great dissention among geologists and geophysicists, because no large mechanical deformation of the solid Earth appeared to be necessary to explain it. It is supposed that the equatorial bulge is self-adjusting to wherever the equator happens to be and that the bulge offers no serious impediment to polar wander. Analysis of satellite orbits has now given a very precise estimate of the ellipticity of the Earth (1/298·25). which is significantly greater than the equilibrium ellipticity (1/300·0), which the Earth would have if all of its internal layers were in hydrostatic equilibrium. It was for some time supposed that the excess ellipticity represented a fossil bulge, corresponding to the equilibrium ellipticity 10 million years ago, and that it demonstrated strength and rigidity of the lower mantle too great to allow spontaneous readjustment of the equatorial bulge. However, the ellipticity excess is now recognised as being due to inhomogeneity of the earth's internal distribution of mass which actually controls the axis of rotation*. Thus it appears that a relatively minor redistribution of mass within the earth could so change the

* See P. Goldreich and A. Toomre. Some remarks on polar wandering. *Journal of Geophysical Research*, Vol. 74, p. 4555 (1969).

balance as to cause polar wandering through a large angle. Precisely this effect is attributed to the process of upper mantle convection which is responsible for continental drift, so that although they may be observationally distinguishable, at least in principle, polar wander and continental drift appear to have a common ultimate cause. Nevertheless polar wander has never been the subject of controversy as heated as that aroused by continental drift for nearly four decades to about 1960. Anti-drifters are still occasionally encountered as fossil relics of that era.

Attempts have been made to explain differences in polar wander curves as a predominance of the non-dipole field over most of geological time. They assume that the dipole character of the Earth's field over the past 10 million years or so is characteristic only of that period and that in the more remote past the magnetic fields of the continents were more or less independent. There are several overwhelming objections to this. Firstly measurements on rocks from widely spaced sites on single continents indicate that the field in the remote past was that of a geocentric dipole and not some more localized feature. Secondly the pole for India appears to have moved so much relative to that for Eurasia that it is virtually impossible to conceive of a field configuration which will account for them both without assuming that India was once widely separated from Asia. Thirdly reversals in the past are known to be correlated between continents; for instance in the Permian period the field was everywhere 'reversed' (that is relative to the simple polar displacement of the continuous polar wander curve) whereas in the Carboniferous and Triassic, which precede and follow the Permian, alternations of field direction were frequent. The field was therefore synchronized all over the Earth. Finally the non-dipole field is known to vary so rapidly when considered on the geological time scale, that if it were dominant in the past then no coherent results could have come from palaeomagnetism at all. Since continental drift is so clearly indicated it is of interest to consider why it was discounted for many years.

The first postulated mechanism of continental drift appealed to the observation, based on gravity and seismic data, that the continental masses are composed of relatively light, granitic (silica rich) rocks, which appear to be floating on the denser underlying rocks in the manner of icebergs on the sea. The ocean floors, below a thin layer of sediments and about 5 km of basaltic rock are composed of the denser ultra-basic material. The continents are in approximate equilibrium, having depths of 30 to 40 km, much greater than their heights; also there are deeper than average 'roots' underneath mountain ranges. It was therefore supposed that the continents could drift as rafts on the Earth's mantle, which, although certainly solid for most purposes, behaved as a fluid when subjected to stresses of geological duration. Although this much was reasonable on the basis of existing evidence, the forces which were invoked to move continents through the mantle material were uncomfortably small. If the mantle is to be regarded as a fluid, a very high viscosity must be assigned to it. Shear waves are transmitted through the mantle with only moderate attenuation and their speed shows that the rigidity of the mantle is actually higher than that of materials at ordinary pressures. Theoretical seismology demanded that continental drift be ruled out as a mathematical absurdity. The arguments of an 'exact' science carried more weight than the qualitative evidence in favour of drift, weakened as it was by differences of opinion about the details of where continents had drifted.

Now that the consideration of drift has been revived by palaeomagnetism, it is explained in terms of a completely different and much more fundamental mechanism. The evidence is now very strong indeed that, in spite of its solid character, the mantle is undergoing convection, and that the continents are merely carried about as surface features on the convective cells. The mantle convection is not necessarily purely thermal, but may well be driven by chemical differences also. This is apparent from the fact that the oldest known igneous rocks occur in continental cores (shield areas) which are presumed to have been the original proto-continents and which have built up to their present sizes by successive additions of progressively younger rocks. The relatively light granitic rocks have separated from the mantle during the convective process and remained at the top as a continental scum.

Since convective stresses can be quite large and the rate of convection required is extremely slow (a few centimetres per year in cells thousands of kilometres in extent) there is no difficulty in accounting at the same time for the solid behaviour of the mantle in transmitting seismic waves. Agreement has not been reached on the details of convection but the principle cannot now be seriously disputed. Mechanical and thermal considerations both point to a convective pattern confined to the upper 700 or so kilometres of the mantle where the temperature gradient is greatest and mechanical strength least.

The magnetism of meteorites

Meteorites provide us with samples of the materials which formed the inner four (terrestrial) planets; their average composition coincides reasonably with the estimated average composition of the Earth. Elucidation of the histories of the meteorites is a vital chapter in the study of the origin of the solar system as a whole. Many of the meteorites are fragments of much larger bodies, probably of asteroidal sizes, up to a few hundred kilometres in diameter.

The meteorites are of several types. The irons which are mainly iron but with a significant fraction of nickel and smaller proportions of other elements, have compositions which are very suggestive indeed of a planetary core. They represent a proportion of meteorite total mass similar to but slightly smaller than the Earth's core as a fraction of the whole Earth. The chondrites form the commonest class. They vary in composition but generally contain some iron in a predominantly silicate matrix, with chondrules or spherical grains of glassy structure, which indicate very rapid crystallization. Their detailed history is by no means clear, but the overall composition is suggestive of a planetary mantle with some iron which had not fallen into the core. The existence of high pressure phases and very extended exsolution structures in both the iron and chondritic types indicates that they originated in bodies hundreds of kilometres in diameter.

Magnetic experiments on chondrites show that many of them have stable thermoremanences induced by fields between 0·1 and 1 oersted.* It has also been demonstrated conclusively that although the outer layers are violently heated by atmospheric friction the heating is so rapid that the interiors are still cool when the meteorites arrive on the Earth. The thermoremanences can therefore reasonably be attributed to an extra-terrestrial cause. The simplest conclusion, and one which has excited some interest, is that the chondrites originated as parts of a planet which had a magnetic field and therefore a well-developed core (now the iron meteorites). While this is a possible explanation it is unwise to place too much reliance upon it. It is not easy to see how almost the entire mantle of even a small planet could have cooled below 600°C while the core remained molten. The obvious alternative explanation is that the chondritic remanence was induced by a very extensive solar magnetic field early in the development of the solar system. The problem of meteorite magnetism thus hangs very closely on the origin of the solar system.

* F. D. Stacey, Palaeomagnetism of meteorites. *International Dictionary of Geophysics*, p. 1141 (Pergamon, Oxford; 1967).

4.3 The Magnetosphere and the Radiation Belts

The extent of the Earth's field in space

The magnetic field of the Earth is similar to that of a dipole which extends outwards into space with intensity varying as the inverse cube, r^{-3}, of the distance r from the centre. The smaller non-dipole components of the field may be represented as the field of a series of higher multipoles, a quadrupole (two opposite dipoles), an octopole, etc. These higher harmonics of the field decrease more strongly in intensity with distance outwards, as r^{-4}, r^{-5} etc., so that the further one goes from the Earth the more nearly correct it becomes to assume that the field is that of a simple dipole. It might, therefore, appear that the Earth's field extends to infinity in all directions, with strength decreasing according to the dipole law of force, and this could be true if interplanetary space were completely empty. But it is not. It is occupied by a very tenuous ionized gas which is an electrical conductor and therefore has a profound effect upon the field. A highly ionized, conducting gas is termed a *plasma* and it is convenient to use this word here.

The geomagnetic field rotates with the Earth, and the field lines therefore sweep through space at a speed which increases proportionately to their distance from the Earth. As was discussed on page 155, the movement of a magnetic field through an electrical conductor induces an electrical current whose interaction with the field has two effects:

(1) it exerts a force on the conductor which tends to make it move with the field, and
(2) it opposes the penetration of the conductor by the field.

Both of these effects are evident in the interaction of the geomagnetic field with the interplanetary plasma. In regions relatively near to the Earth, that is within about 10 Earth radii, the field is strong enough to make the plasma rotate with it. At greater distances, where the field is weaker, the plasma does not rotate with the Earth, but instead it prevents the penetration of the field. The field is thus confined to a limited volume, the *magnetosphere*; inside it the plasma is controlled by the field and moves with the Earth; the plasma outside is independent of the Earth. In the language of plasma physics the magnetosphere is a giant magnetic bottle; it contains the very extensive outer part of the atmosphere, the gaseous envelope which travels with the Earth in its orbit around the sun; most of it is very tenuous indeed. The magnetospheric shape is determined by the fact that the interplanetary plasma is not stationary but is moving radially outwards from the Sun (the solar wind) but the existence of the magnetosphere itself does not depend upon this motion. The rotation of the Earth and its motion about the Sun would produce a magnetospheric 'bottle' even in stationary plasma. The physics of the magnetosphere has developed dramatically in the past decade and there is now a vast literature*.

The magnetospheric plasma is highly ionized, that is it contains very few neutral atoms and consists mostly of electrons and positive ions (largely protons with some helium nucleii and possibly a few heavier ions). Its lower boundary is the *ionosphere*, which extends from about 60 km above the earth's surface to about 300 km. In this region the density is greater than in the outer parts of the magnetosphere and the ionization is less complete. The proportion of neutral atoms present increases downwards from nearly zero at several hundred kilometres, to nearly unity below 60 km. The composition also changes with decreasing height, being largely hydrogen in the magnetosphere and a nitrogen-oxygen mixture in the dense part of the atmosphere. In spite of the presence of neutral atoms the high concentration of ions in the ionsophere makes it a better electrical conductor than the outer magnetosphere. Magnetospheric electrical currents excited by geomagnetic disturbances are therefore particularly strong in the ionosphere. However the ionosphere itself plays only a secondary role in the disturbances and is not a primary cause.

The solar wind

The Sun emits a continuous stream of charged particles comprising equal numbers of protons and electrons. This is known as the 'solar wind', an electrically neutral plasma 'blowing' out from the Sun in all directions. Its effect upon the magnetosphere may be visualized as that of an ordinary wind blowing on a flimsy and very elastic balloon which is held in place by light elastic threads to a ball (the earth) inside it. That day-to-day variations in the geomagnetic field are intimately connected with variations in solar emission has been known for many years. Direct observations of the solar wind by deep space probes have provided data on particle velocities and densities and confirmed that the essential features of their interaction with the geomagnetic field are reasonably well understood.

Valuable observations of the solar wind were relayed by Mariner II on its long flight to Venus. The particle velocity is normally about 500 km/sec, with occasional peaks exceeding 1,200 km/sec. This velocity is much

* It is extensively reviewed in a collection of papers edited by S. Matsushita and W. H. Campbell, *Physics of Geomagnetic Phenomena*, 2 volumes (Academic Press, New York; 1967).

greater than that of any space probe, so that motion of the probe could be neglected in analysing the motion of particles impinging on it. The particle flux (at the distance from the Sun of the Earth's orbit) averaged 5×10^8 per square centimetre per second, so that the particle density in space was only about 10 per cubic centimetre. Nevertheless, because of the enormous extent of the solar wind, interactions of particles with one another are quite significant; they do not have to be considered individually but can legitimately be treated as a plasma. These figures are representative of the quiet-time solar wind, which probably varies by a factor of two or perhaps four during the 11-year solar cycle, quite apart from the hour-to-hour and day-to-day fluctuations. The particle emissions responsible for magnetic storms are much more energetic; consideration of their effects is reserved to Section 4.4.

For equal proton and electron velocities, most of the solar wind energy is kinetic energy of the more massive protons. The solar wind velocity corresponds to an energy of about 1 200 electron volts (e.v.)*, for protons but only about 1 e.v. for electrons, and the energy of recombination to form hydrogen is only 1 per cent of the proton kinetic energy. The total energy of the solar wind reaching the earth is therefore simply the product of proton energy and particle flux and works out to about 1 erg per square centimetre per second. In spite of the impressive numerical values for particle velocity and particle flux, the actual energy flux in the solar wind is quite small, as can be seen by comparing it with the energy flux of electromagnetic radiation. The solar constant, or energy flux, received at the Earth from the whole spectrum of light emitted by the Sun is 2 calories per square centimetre per minute or $1{\cdot}4 \times 10^6$ ergs per square centimetre per second, a million times greater than the energy of the particle flux. However, the radiation is intercepted only by the solid earth (and to a limited extent by the dense part of the atmosphere), whereas the solar wind is intercepted by the whole magnetosphere which has a cross-section 100 times as great, so that the factor by which radiation energy exceeds the total particle energy intercepted by the Earth and the magnetosphere is only 10^4. This is still sufficient to show that large fluctuations in solar particle emission have very little effect upon the total energy received by the Earth and explains why there is no obvious dependence of meteorological conditions upon solar particle emissions.

In any small volume of the solar wind plasma the proton velocities are not all equal. There is a distribution of velocities about the mean and this has been represented as a proton 'temperature' one to a few hundred thousand degrees centigrade, which amounts to a velocity variation of about 10 per cent. This randomness in proton velocity has nothing to do with the variations in mean velocity which are evidently a large scale plasma

* An electron volt is the kinetic energy acquired by an electron (or any particle with the same magnitude of charge) which is accelerated through a potential difference of 1 volt. 1 e.v. $= 1{\cdot}6 \times 10^{-12}$ erg.

effect, reflected also in the variable strength of the interplanetary magnetic field.

The magnetic field in interplanetary space has a strength of a few gammas,* generally directed at about 45° to the earth-sun line, in the manner of streaming of field lines from the rotating sun. The field variations observed by Mariner II had periods from 40 seconds upwards. Since this reflects the scale of the irregularities in the solar plasma, the irregularities are from 20 000 km upwards in extent, probably with no upper limit short of the earth-sun distance (150×10^6 km). The fact that irregularities in the plasma, 20 000 km in extent can survive a three to four day transit from the sun while the protons have a random velocity amounting to 50 km per second which would smooth them out in a matter of minutes, points to the localized magnetic control of the plasma. We may visualize the solar wind as a stream of water which has, in addition to its clearly observed steady flow, a superimposed pattern of irregularities and eddies. The eddies are controlled and stabilized by local magnetic fields with associated electric currents arising from the relative motion of protons and electrons. The fields are generated by the vortex motion itself and act as magnetic bottles to conserve the volumes of plasma which produce them. Evidently, the plasma density is insufficient to support magnetic fields stronger than a few gammas, so that vortices smaller than about 20 000 km cannot be stabilized.

As the solar wind meets the geomagnetic field there is a mutual repulsion of the wind and field. The magnetosphere is compressed on the side facing the sun and extended on the opposite side while the solar plasma is deflected around the magnetosphere. The resulting shape of the magnetospheric boundary is shown in cross-section in Figure 52, the sun-ward side being much more clearly delineated than the long 'tail'. In 'quiet' solar conditions, that is when the Sun emits the normal solar wind but not the more energetic particles which cause magnetic storms, the boundary of the magnetosphere on the Sun-ward side is at about ten Earth radii (60 000–70 000 km). At this distance the strength of dipole field is smaller by a factor 10^3 than its value at the surface (one Earth radius). This means that at the equator on the Sun-ward side, the field would be about 30 gammas. The effect of compression by the solar wind (plus the 'ring current' discussed in the following section) is to increase this to about 50 gammas at the beginning of a turbulent transition zone, in which the field has an average value of about 50 gammas but fluctuates in magnitude and direction. Outside the magnetospheric boundary there is a shock wave in the solar wind, owing to the fact that it is streaming against the magnetospheric bottle at a velocity several times the speed of sound in the medium.

The physical process of magnetospheric compression by the solar wind can be described, in part, as a separation of the solar protons and electrons, which are deflected as they meet the field. The continuous process of charge separation constitutes a current in the magnetospheric surface. Its form is

* 1 gamma = 10^{-5} oersted.

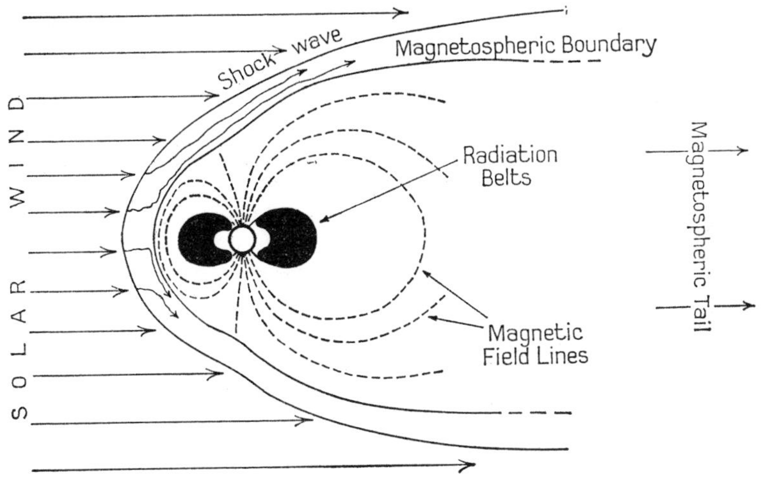

FIG 52 A diagrammatic cross-section of the magnetosphere, showing the effect of the solar wind on its shape, and the positions of the radiation belts

such that it cancels the dipole field outside the surface and enhances it within the surface. 'Friction' of the streaming solar wind against the magnetospheric boundary is one mechanism by which energy is fed to the magnetosphere by the solar wind.

The 'leeward' side of the magnetosphere is more variable as well as being more uncertain in its form than is the Sun-ward side and, although the magnetospheric surface presumably closes somewhere in the region 50 to 60 Earth radii, the closure has not been observed. Its remoteness from the Earth makes it more difficult to study with satellites than is the sunward side. It fluctuates rapidly, behaving rather like a flag flapping in the wind. Possibly there is no distinct closure but the field behaves rather like a flame with sections breaking away and disappearing into space as eddies in the solar plasma.

The radiation belts and the ring current

In the preceding sections it was sufficient to treat the solar wind and magnetospheric particles collectively as 'plasma'. This is no longer a satisfactory approach when we have to consider the motions of the very energetic particles within the magnetosphere. Instead it is necessary to consider the particles as individuals. Each moving charged particle may be treated as an electric current which experiences a lateral force ($\boldsymbol{G}$) from its interaction with the magnetic field ($\boldsymbol{F}$). In vector notation

$$\boldsymbol{G} = e\boldsymbol{v} \times \boldsymbol{F},$$

where $\boldsymbol{v}$ is the particle velocity and e its charge. Considering first a uniform field, $\boldsymbol{v}$ may be resolved into components v_p parallel to $\boldsymbol{F}$ and v_n normal to $\boldsymbol{F}$. No force results from the parallel motion which is therefore unaffected by the field. The motion normal to $\boldsymbol{F}$ gives rise to a force perpendicular to both $\boldsymbol{F}$ and v_n which thus causes it to describe a circular path of radius (mv_n/eF) if the particle has a mass m. The combination of circular motion with speed v_n about $\boldsymbol{F}$ and a steady motion v_p parallel to $\boldsymbol{F}$ results in a spiral path about the lines of magnetic force. This is the essential process of particle trapping. Energetic particles in a uniform field are confined to a particular tube of force whose dimensions are determined by the component of velocity perpendicular to the field.

Particle motion in a dipole field, such as that of the Earth, is more complicated and two further effects have to be considered, resulting from

(1) the curvature of the lines of force and their concentration towards the poles, and
(2) the fact that the field intensity decreases outwards.

However, provided that the width of a spiral is very small compared with its distance from the centre of the Earth, the spiral path is a satisfactory first approximation and the dipole field effects can be treated as perturbations of the spiral. This is a valid assumption for virtually all electron energies encountered and for protons up to about 10^7 electron volts. The motion of more energetic protons must be qualitatively similar but is not subject to a simple analysis.

Since we are considering spirals small compared with the scale of the dipole field and the electromagnetic force experienced by a particle is perpendicular to the field, its spiral path is bent around to follow the field lines. At the same time as curving towards a pole the field lines converge and the field becomes more intense, causing the particle to describe a tighter spiral as it approaches the pole. Thus it is still confined within a particular (converging) tube of magnetic force. The convergence has a further effect on the particle motion which is more readily visualized by analogy with a light beam projected into a reflecting funnel of slight convergence (Figure 53). At each successive reflection the angle of incidence of the light is changed so that instead of continuing indefinitely down the funnel it is reflected back out of the opening. In a very similar manner a particle trapped in a particular tube of magnetic force is prevented from approaching the pole and is 'reflected' back towards the equator by the convergence of the field. The point of its closest approach to the pole is the 'mirror' point. A particle reflected back along a tube of force from the mirror point in one hemisphere crosses the equatorial plane and makes a similar approach to the opposite pole and is then reflected from its mirror point in that hemisphere. Particles 'trapped' in the geomagnetic field oscillate indefinitely between their two mirror points. The regions of the magnetosphere in which intense fluxes of trapped protons and electrons

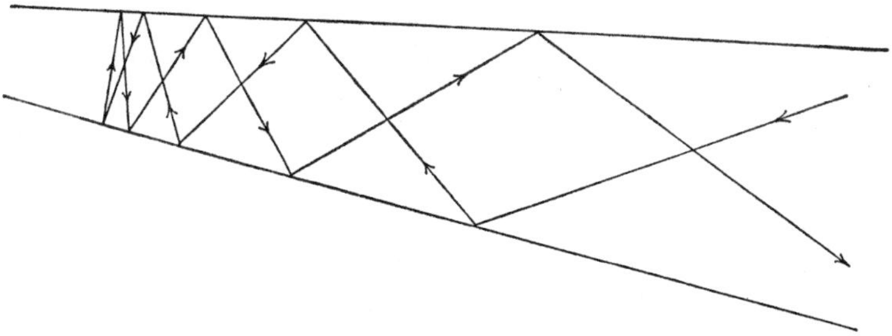

FIG 53 Light beam analogy to 'mirroring' of trapped particles in the geomagnetic field. The light beam penetrates the reflecting funnel to a distance determined by its initial angle of incidence

have been found have been named Van Allen belts after their principal discoverer Professor James A. Van Allen. Although the discovery had been anticipated theoretically, when the particle radiation was observed by satellites, its intensity came as a surprise.

An important additional feature of the radiation belts, particularly the outer part, arises from the radial gradient of the geomagnetic field. The curvature of the path of a particle spiralling in the field is determined by the intensity of the field; the stronger the field, the tighter is the spiral. Thus in the Earth's field, the curvatures of the inner and outer parts of a spiral differ slightly because the inner part, being nearer to the Earth, is in a stronger field; this imparts a slow lateral drift to the spiral path, in the manner of Figure 54. Since the charges of protons and electrons are opposite in sign, they spiral in opposite directions and the drift direction is also opposite, westwards for protons and eastwards for electrons. The drifts

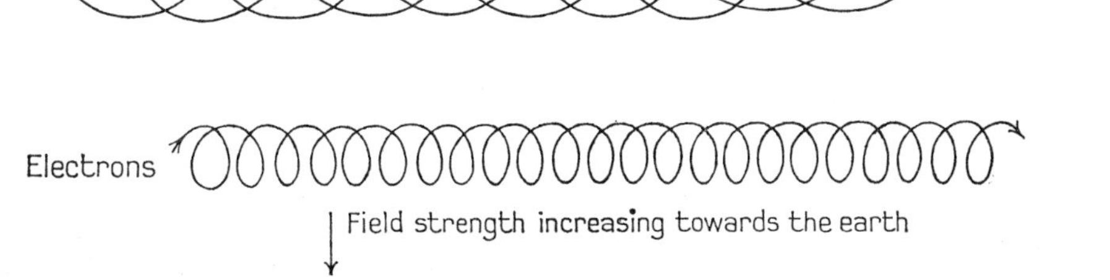

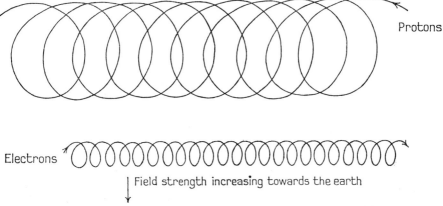

FIG 54 Lateral drift of the spiral paths of radiation belt particles, resulting from radial variation in the strength of the Earth's field. The particle motion is here viewed along the lines of magnetic force

of both protons and electrons thus contribute in the same sense to a westward electric current encircling the Earth. This is the geomagnetic ring current, centred at about 6 Earth radii in the equatorial plane and rather closer at higher latitudes. The existence of such a current was suspected half a century ago from magnetic storm data, but the motion of the particles which give rise to it was first suggested by S. F. Singer, two years before the first reports of observations of the radiation belts in 1959.

During magnetic storms the intensity of the ring current fluctuates violently, but in magnetically quiet times it appears to be reasonably steady. The field which it produces at the Earth is then of the order 30 gammas, being approximately uniform and parallel to the Earth's magnetic axis over the surface of the Earth (which is small by comparison with the dimensions of the radiation belts). At the equator the field of the ring current is opposite to the main field and at the magnetic poles it reinforces the main field, as will be apparent from the geometry of the dipole field in Figure 40. The field of the ring current immediately outside the outer part of the radiation belt, being opposite in direction to the field inside, reinforces the main field and thus pushes the magnetospheric boundary slightly further away from the Earth than it would be in the absence of the current. At the radial distance from the Earth of the outer part of the radiation belt the field of the ring current is more than 10 per cent of the dipole field, so that an analysis of the particle motion in the dipole field only can be appreciably in error; the contribution to the field of the ring current itself has also to be taken into account. During storm-time enhancement of the ring current the field within the outer belt may be locally annulled and this appears to be of particular interest in connection with the cause of aurorae (section 4.3).

The lifetimes of geomagnetically trapped particles are highly variable. Under magnetically quiet conditions some particles appear to remain trapped for months or even years; the most striking evidence of this was the prolonged effect of the particles injected into the radiation belts by the 'Starfish' explosion of a 1·4 megation nuclear device at an altitude of 400 km, over Johnston Island in the Pacific in July 1963. On the other hand the massive numbers of particles dumped into the ionosphere at the auroral zones during magnetic disturbances can have had only transient existence as energetic trapped particles.

In spite of the intensity of the particle flux, the actual particle density in the belts is sufficiently low for collisions to be rare, and, even so, many of the collisions which occur could not change the particle velocities sufficiently to deflect them into non-trapping orbits. The lifetime of a particle in a radiation belt is therefore determined principally by the height of its mirror points above the dense part of the atmosphere. A particle which is reflected from a level in or near to the ionosphere stands a much greater chance of being intercepted by gas atoms and being lost to the belt than a particle whose mirror points are at 1,000 km or more. The height of the mirror points are determined by the distance from the Earth at which the particle crosses the equatorial plane and by its *pitch angle* at the equator, θ_e, that is the angle which its path makes with the field direction. In terms of the velocity components v_p and v_n (page 180):

$$\tan \theta_e = \frac{v_n}{v_p}.$$

Calculations show that for particles crossing the equatorial plane at 6 Earth radii, only those particles with pitch angles smaller than about 3° can reach the atmosphere, whereas, of the particles which cross the equatorial plane at 3 Earth radii, those with pitch angles smaller than 9° reach the atmosphere.

The mechanisms of loss and replenishment of trapped particles in the radiation belts are major unsolved problems. There are several theories, but none of them appears to be adequate alone and it is likely that the several mechanisms of regeneration all operate, supplying particles of various energies to different parts of the belt. It appears necessary to suppose that at least some solar particles enter the magnetosphere. A few highly energetic solar (cosmic ray) protons can do so directly but their numbers are too small to account for the population of charged particles; the majority must enter at much lower energies through instabilities in the magnetospheric surface at the neutral points, where the lines of force from the poles meet the surface and, probably more importantly, through the tail. However, the energies of solar wind particles are too low for them to become radiation belt particles directly and an acceleration mechanism by hydromagnetic waves must be invoked. Another important mechanism for the production of energetic particles in the radiation belts is the decay in situ of cosmic ray neutrons to protons and electrons. The neutrons are produced by interaction of the primary cosmic ray particles with the Earth or the atmosphere and would escape from the Earth altogether except that some of them happen to disintegrate radioactively while still within the magnetosphere.

Cosmic rays are very energetic charged particles, which arrive at the Earth from space, some with energies sufficient to penetrate the magnetosphere with relatively little deflection and reach the Earth. There appear to be two distinct sources. Very fast protons from the sun can reach the earth as cosmic rays; even more energetic particles which include many heavier atomic nucleii as well as the more abundant protons arrive at the Earth from remote cosmic sources which are distributed equally in all directions.

The principles of the interaction of cosmic ray particles with the geomagnetic field are the same as for the trapped particles of the radiation belts. The magnitude of their deflection by the field is determined by a parameter, known as magnetic rigidity R_m. For a particle with velocity v and charge-to-mass ratio e/m, $R_m = v(m/e)$. Rigidity, or momentum-to-charge ratio, is a quantitative measure of the ability of a particle to penetrate the geomagnetic field. It is immediately apparent that the electron component

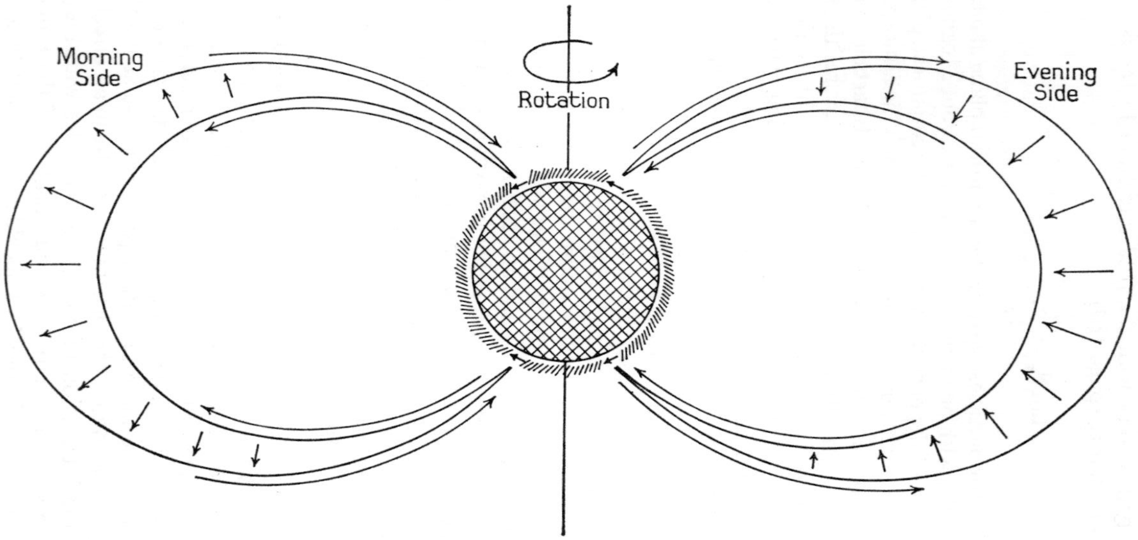

FIG 55 A mechanism for the generation of electric currents in the auroral zone ionosphere by tidal drift of plasma through a radiation belt in which protons have higher momenta than electrons. The net radial current flow through the belt is compensated by flow along the geomagnetic field lines to the ionosphere. Arrows indicate direction of current flow, being opposite to the electron motion

of cosmic radiation is much less effective in reaching the Earth than is the proton component. Further, protons with moderate cosmic ray energies can penetrate more easily to high geomagnetic latitudes than to low latitudes. In an extreme case we can imagine a particle reaching one of the magnetic poles by travelling *along* a line of force all the way. There are more lines of force to be *crossed* in reaching the magnetic equator than anywhere else. It is therefore observed that the cosmic ray particles which arrive at low latitudes do not include those with low rigidity, which can nevertheless reach the earth at high latitudes. For each geomagnetic latitude there is a *cut-off rigidity* below which cosmic ray particles are not observed.

The magnetosphere in motion

The gross asymmetry of the magnetosphere is apparent in Figure 55. The field is compressed on the Sun-ward side and drawn out into a long tail on the opposite side. Since the field rotates with the earth, it is evident that there is a diurnal compression and expansion of the field, in other words a magnetospheric tide, which is apparent as a small diurnal variation of the local geomagnetic field. Three effects can be distinguished. Firstly the simple fact of compression causes a general enhancement of the field on the sunlit side of the Earth. Secondly a frictional drag is exerted on the magnetospheric boundary by the solar wind, drawing plasma from the Sun-ward side to the tail, from which it can return through the core of the tail to complete the cycle.* Thirdly the relative motions of plasma and field lines cause charge separation, generating electrostatic fields and currents, the most striking of which are in the auroral zone ionosphere. The magnitude of the magnetic diurnal variation observed at the surface of the earth is 10 to 20 gammas, which is, as expected, comparable to the strength of the dipole field at the magnetospheric boundary, where the interactions causing the variation originate.

A feature of the charge separation effect, to which attention has been drawn by J. A. Fejer, is that the higher the momentum of an ion in the magnetosphere, the less readily it responds to the changing magnetic field. Thus the low energy plasma participates in the tide, but the energetic particles in the radiation belts do so much less, if at all. Further, electrons move more readily than protons. The tidal movement of plasma with the diurnal pulsation of geomagnetic field lines therefore carries a net negative charge through the radiation belts, in the manner of Figure 55, that is an electric current flows radially through the belt. The current flow will be partially inhibited by the electric field built up by the space charges of surplus protons and electrons on the inner and outer surfaces of the belt, but there is a path by which the current can complete a circuit and thus be maintained. Circumferential flow of current around the surface of the belt

* This process has been termed 'magnetospheric convection'.

from the evening side of the magnetosphere to the morning side, where the electric polarity is opposite (the morning side is being compressed while the evening side is dilating) must be ruled out because it requires particle motion across the field lines. Particle motion along the field lines is allowed and the pattern of current flow is therefore along the field lines to the arctic and antarctic ionosphere, as in Figure 55. Current flow from the inside surface to the outside surface of the belt or vice versa is completed through the ionosphere in the auroral zones at the 'ends' of the belt. In addition to being the shortest path for completion of the current circuit, this is also the path of greatest conductivity.

In the auroral zone ionosphere the magnetospheric current flow is across the field lines and this causes a strong *Hall effect*. The required current flow is in a north-south sense and since the Earth's field is nearly vertical (i.e., radial) at high latitude, it deflects the current to a substantially east-west direction. As in Figure 56 the direction of this east-west current is opposite in the morning and evening hemispheres, continuity being maintained by 'leakage' across the polar cap and probably also via low latitudes. The auroral zone current is known as the *auroral electrojet* by analogy with jet streams in meteorology. It is much more intense than the current flowing in any other part of the ionosphere. The observed auroral electrojet current system differs from that obtained in the above simple model, being ad-

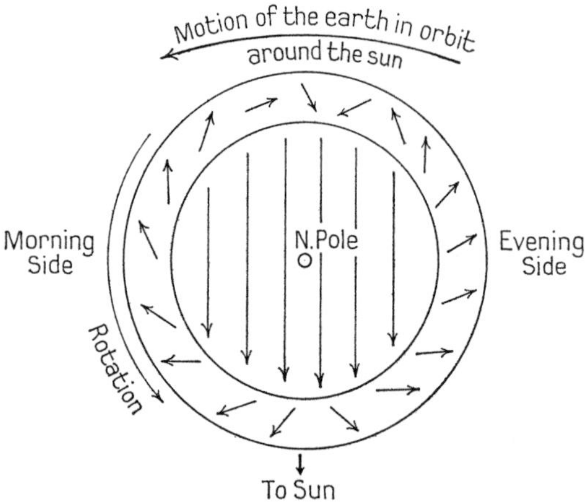

FIG 56 View from above the pole of the directions of current flow in the ionosphere of the north polar region, due to currents generated in the radiation belt by the mechanism illustrated in Figure 55. The circumferential flow is a consequence of the Hall effect. Note that current enters and leaves the ionosphere from the region of the radiation belts, so that lines of current flow do not close in a two-dimensional picture of the ionosphere, although in many presentations they are shown as doing so

vanced in phase, that is, the current pattern is rotated about 30° clockwise relative to that shown in Figure 56.

Resonant effects and micropulsations

Various types of small, rapid fluctuations of the Earth's field, known as *micropulsations*, are due to oscillations in the magnetosphere, but the mechanisms for their generation are not well understood. This is because the internal structure of the magnetosphere is rather like a complex system of interconnected springs and suspended weights. If such a system is disturbed in one place the disturbance spreads, causing various motions which may have periodic tendencies. But in the case of micropulsations, the linkages are not observable and one can only make guesses at the ultimate cause of a disturbance, so that it is very difficult to deduce either the cause or the linkages from observations of the resulting disturbances. The study of micropulsations has developed almost entirely in a phenomenological way, with the classification and reclassification into an increasing multiplicity of types of micropulsation, which in many cases are not demonstrably distinct or different effects.

Disturbances in the magnetosphere are propagated as hydromagnetic (or magnetohydrodynamic) waves of two types, longitudinal and transverse. They are waves in the magnetic lines of force in which the conducting plasma moves with the field lines and controls their motion. In mechanical terms the field lines are elastic and the plasma, which is 'frozen' to them, provides the inertia of the system. It is convenient to think in terms of classical tubes of magnetic force. Then the longitudinal waves are like kinks which travel along the tubes of force and transverse waves are compressions which are transferred between tubes and thus propagate across them. All of the tubes of force are 'fixed' in the Earth by the relatively high conductivity at a depth of a few hundred kilometres. A wave reaching the earth along a tube of force is therefore reflected, and, if its half period is a submultiple of the time taken to travel between the *conjugate points* at the opposite ends of the tube, a standing wave is set up. Simultaneous records of micropulsation activity at conjugate points have verified that both out-of-phase and in-phase micropulsations occur at the two stations, corresponding to waves which are respectively odd and even sub-multiples of the line of force joining the points. Other possibilities for the establishment of hydromagnetic waves include resonance between the Earth and the magnetospheric boundary and around lines of latitude in the magnetosphere. In many cases several types of waves are probably excited at the same time.

4.4 Magnetic Storms and Associated Effects

Solar activity and magnetic disturbances—the 11-year cycle

The relationship between magnetic disturbances and solar activity has been known for many years and in the early 1930's S. Chapman and V. C. A. Ferraro explained the connection in terms of neutral streams of charged particles projected towards the Earth from the Sun. The structure of the magnetosphere and the effects occurring within it, as outlined in Section 4.3 are, by comparison very recent discoveries. However they make it possible to give a reasonably simple and logical account of the earlier discoveries. The relationship between solar activity and magnetic disturbances can be seen in individual magnetic storms, but is far more obvious when time-averages of magnetic and solar activity are compared. Particle emissions from the Sun have a general association with sunspots. Although the occurrence of sunspots does not itself indicate these emissions, when the sun is very spotty there is a much higher probability of energetic particle emissions. Thus sunspots, being obvious features of the solar surface, have for many years served as indicators of surface activity in general. The most used index of the Sun's spottedness is the Zurich sunspot number, r, computed daily at the Swiss Federal Observatory from observations made all over the world. It is defined by the formula:

$$r = k(f + 10g)$$

where g is the number of sunspot groups and f is the total number of spots in the visible hemisphere. k is a numerical factor, slightly different values being applied to the data from the various contributing observatories to compensate for local differences in instrumental efficiency and in the somewhat subjective estimation of what constitutes a 'group' or even an individual spot. The index is a reliable and very useful one in spite of appearing to have a rather arbitrary definition.

Another useful measure of sun-spottedness which has the advantage that no arbitrary definition is involved is the total sunspot area, the fraction of the visible hemisphere of the Sun occupied by sunspots, which is tabulated regularly by the Royal Greenwich Observatory. Annual means of sunspot area are reproduced from the Greenwich data in Figure 57 and compared with annual means of the daily range in geomagnetic declination. These curves show clearly the 11-year cycle of solar and magnetic activity. The

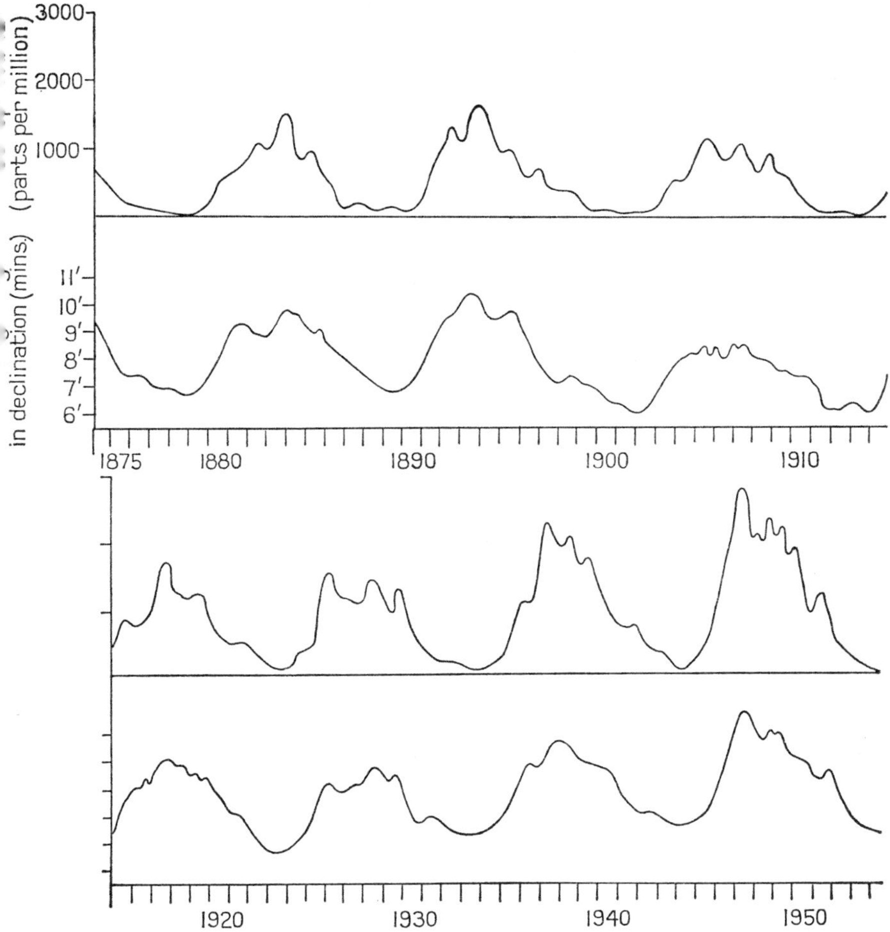

FIG 57 Twelve month running means of sunspot area (parts per million of the visible hemisphere) and the daily range of geomagnetic declination (minutes of arc). (Curves reproduced by permission of the Controller of Her Majesty's Stationery Office and the Astronomer Royal from 'Sunspot and Geomagnetic Storm Data, 1874–1954', by the Royal Greenwich Observatory)

correspondence between the two leaves no doubt that the Sun is responsible for geomagnetic disturbances.

The nature and underlying cause of the solar cycle are of considerable interest but remain largely speculative at present. The outstanding observations of H. W. and H. D. Babcock have shown that the complete cycle is 22 years, with reversals of the solar magnetic field occurring at about the sunspot maximum every 11 years. It has been known for much longer that there are strong magnetic fields in individual sunspots so that their cause is evidently directly related to the main field of the Sun. At sunspot minimum the main field probably has a reasonable resemblance to that of a dipole but at sunspot maximum, when the field reverses, it is much more complicated and may even have poles temporarily of the same sign in both hemispheres, so that the magnetic flux emerges at low latitudes, apparently providing a radial field which stimulates the generation of sunspots.

The predominant latitude of sunspots varies through the 11-year cycle. They are at all times virtually confined to low solar latitudes, within about 35° of the sun's equator, although rarely on the equator itself. A new cycle of sunspots commences at sunspot minimum with the appearance of spots around 30° (north and south). At maximum activity spots occur with quite a wide range of latitudes centred around 15° (N and S) and as the cycle subsides into the following minimum of activity the spots are grouped around 5°–10°, sometimes appearing at the same time as the beginning of a new cycle at 30°.

The axial rotation of the Sun is readily observed from the transits of sunspots. Although substantial variations occur at all latitudes, the equator is rotating generally more rapidly than the middle latitudes, which may indicate a stronger convective tendency at the middle latitudes. The average period of rotation at sunspot latitudes is about 27 days and this gives rise to a tendency for magnetic disturbances to recur at intervals averaging 27 days, as was shown at the beginning of this century by C. Chree. The fact that 27 (and 54, etc.) day recurrence is noticeable in magnetic data shows not only that magnetic disturbances are due to emissions from limited areas of the Sun's surface, but that an active area survives, at least in some cases, for several solar revolutions.

Sequence of events during a magnetic storm

Even on magnetically 'quiet' days the geomagnetic field is never completely undisturbed. There is always some fluctuation which, may reasonably be ascribed to variations in the solar wind. There are also more disturbed days when fluctuations are quite noticeable on the records of magnetic observatories, but which reveal no particularly characteristic sequence of events. Major disturbances are referred to as magnetic storms, each of which does have a recognizable characteristic sequence or pattern of disturbance. Sometimes several storms will follow one another in rapid succession, so that their effects are superimposed, but even so, their separate onsets can often be distinguished, because their individual pattern is sufficiently well established.

Some magnetic observatories have for many years classified magnetic storms in two types, those which have sudden commencements and those which do not. The ultimate causes of both types can be traced back to the sun, but the mechanisms are somewhat different. Storms without sudden commencements are due to an intensification, but no marked speeding up of the solar wind. Their onsets signal the arrival of solar plasma of greater than normal density which causes an increased compression of the magnetosphere. The stream of dense plasma has no sharp boundary and therefore the magnetospheric compression does not begin suddenly but builds up gradually. The plasma generally originates from a recognizable area of disturbances on the solar surface, such as a group of sunspots, from which it travels radially outwards. The transit time to the Earth is about three days so that magnetic storms (that is those without a sudden commencement) begin about three days after the solar disturbance is observed to cross the centre of the Sun's disc. Since the sun rotates on its axis once every 27 days, it has turned about 40° by the time the dense plasma reaches the Earth. The path of such a plasma stream is indicated in Figure 58.

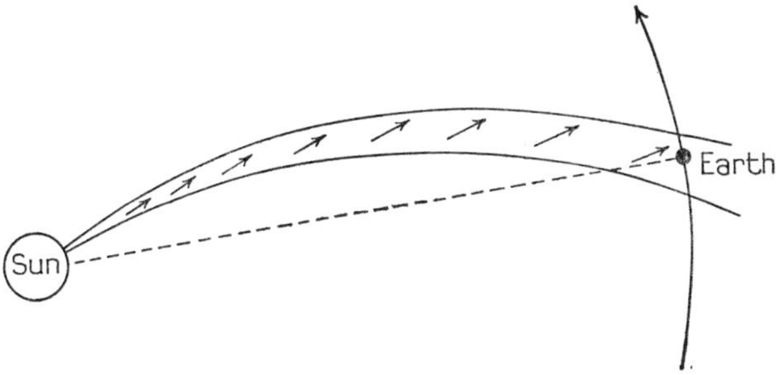

FIG 58 Stream of plasma from a disturbed area of the Sun's surface. The plasma moves radially outwards from the sun but the Sun's rotation gives a curved outline to the stream, similar to that of a jet of water from a rotating nozzle

Occasionally there is a violent eruption on the Sun's surface, generally recognizable as a *solar flare*, which produces not only an intense burst of light, particularly in the ultra-violet, but also high velocity plasma. A solar flare occupies only a small part of the Sun's surface, but its energy output is prodigious. The ultra-violent light received by the Earth increases noticeably and causes enhanced ionization in the ionosphere on the sunlit side of the Earth. Ionospheric currents are consequently increased temporarily and a 'notch' appears in the records of magnetic observatories which happen to be on that side of the earth at the time of the flare. This magnetic effect is known as a 'solar flare effect'. The high velocity plasma is not only several times faster than the normal solar wind (2,000 km/sec or more) but its intensity, at the radius of the Earth's orbit, can be ten times as great as that of the solar wind. Its transit time to the Earth is one day or less and its arrival at the boundary of the magnetosphere causes a sudden compression, which reaches the Earth as a sharp magnetic pulse, the 'sudden commencement' of a magnetic storm. All really great storms are of this type.

The sudden commencement pulse is transmitted from the magnetospheric boundary to the earth as hydromagnetic waves, very similar to those which are responsible for micropulsations (page 187). A transverse wave (lateral compression of the tubes of magnetic force) is propagated across the lines of force mainly to low latitudes and a longitudinal wave or 'kink' travels along the tubes of force to high latitudes. The waves reach the Earth's surface as nearly as can be determined simultaneously everywhere. Their amplitudes are commonly several tens of gammas and not infrequently over 100γ.

All magnetic storms, whether with or without sudden commencements, have an initial phase of continuing magnetospheric compression, during which the strength of the field at the surface, particularly the horizontal component*, is increased. This is followed by a main phase of greater intensity and duration than the initial phase, the main phase being characterized by a diminution in the field strength. The whole sequence of events during a magnetic storm, as revealed by records obtained at magnetic ob-

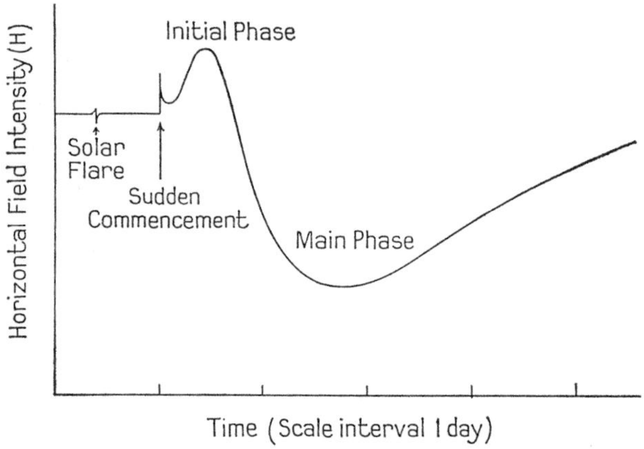

FIG 59 Characteristic sequence of events during a magnetic storm, idealised from records of horizontal magnetic field intensity.

servatories, is shown in Figure 59. This is not an actual record but has been idealized by eliminating complications such as multiplicities of commencement.

During the development of the main phase of a magnetic storm the population of the radiation belt undergoes marked changes. Whereas the quiet time ring current is believed to produce a field of about 30γ at the sur-

* The effect of electrical conduction at depth in the mantle is to increase the horizonta component of a magnetic disturbance and decrease the vertical component.

face of the Earth, the storm-time ring current can produce a field 30 times as great. However the enormous enhancement of the population of the radiation belt appears to leave the quiet time population more or less unaffected. It is mainly the new particles which are dumped into the ionosphere in the auroral zones during the recovery period of a storm. Recovery of the field, which is a measure of the decay of the ring current, takes several days.

Magnetic storms generally have a structure indicating multiple arrivals of fresh solar plasma, noticeable in polar regions as sub-storms. Since the solar plasma is magnetically controlled, carrying its field along with it, we can think of a series of large and energetic plasma vortices, each with its own built-in field and associated internal motion. Such a vortex is likely to be about 10^7 km across, much larger than the whole magnetosphere, which is enveloped within a vortex during the hour or so which it takes to pass the earth. The irregular growth and decay of auroral electrojet currents is in accord with this picture.

Aurorae

Aurorae are a result of the still very imperfectly understood process by which energetic particles enter first the region of the radiation belts and from there the ionosphere in the auroral zones. When the particles are intercepted by the denser air they are gradually slowed down by collisions with neutral atoms and then absorbed. Part of their energy is transferred to the neutral atoms, whose electrons are excited to energy levels from which they can subsequently fall back, producing radiation in the process. Ionization is the limiting case of such excitation. The patterns of light emitted during an auroral display are patterns of particle precipitation. They are sufficiently complex that a really satisfying detailed explanation will probably elude us for many years yet, although the clear association of aurorae with the geomagnetically trapped particles makes it clear that the understanding of aurorae will progress simultaneously with knowledge of the radiation belts.

During aurorae both electrons and protons are precipitated into the ionosphere over a wide latitude range and not merely in the limited zone of visible aurora, although the precipitation is there most intense. The visible effects are associated primarily with the precipitation of electrons in the energy range tens to hundreds of kilo-electron volts. The particle flux is so great that simple dumping of stored particles from an unreplenished radiation belt would in extreme cases deplete the belt in a matter of minutes or less and very much less than the life-time of a storm. A very rapid process of acceleration and dumping of particles must occur, whether or not they are captured from the solar plasma through instabilities in the magnetospheric boundary. However, the mechanism is the subject only of guesses.

Details of the precipitation mechanism are similarly unknown, but observations provide some clues. In particular the zone of precipitation

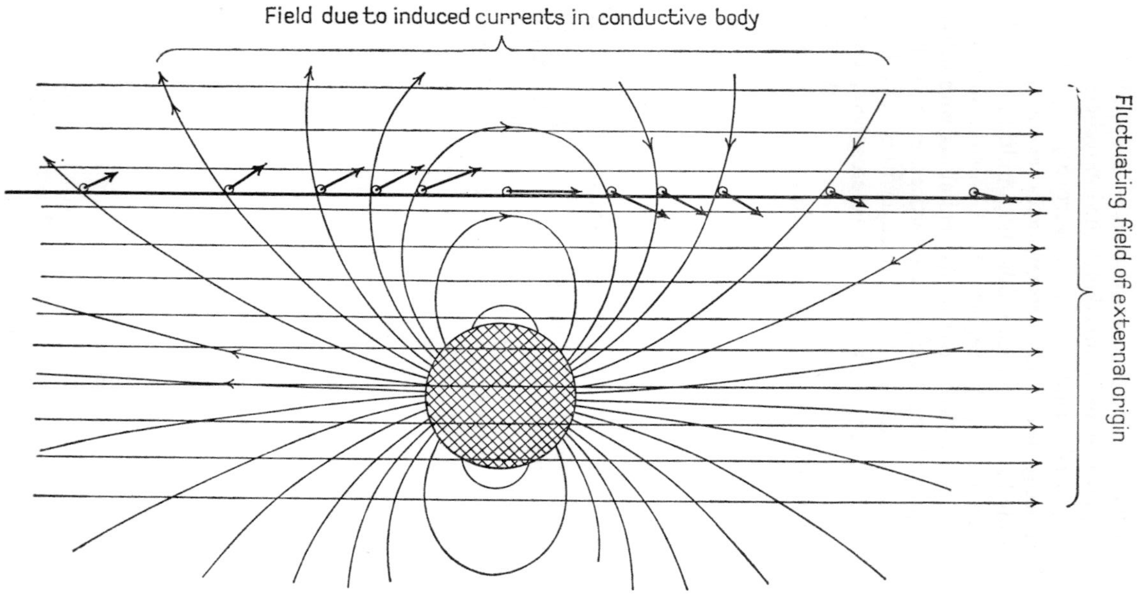

FIG 60 The distortion of a magnetic disturbance in the region of a conducting body. It is assumed that the lines of force of the primary disturbance field would be uniformly horizontal in the absence of the conducting body which is shown shaded. The eddy currents which are induced in the body circulate to produce a secondary field which opposes the horizontal primary field in the body but also extends beyond the body in the manner of the curved field lines. The resulting disturbance vector at any point is the sum of the primary and secondary fields. The vectors at a series of points along a horizontal surface are shown by heavy arrows. Phase differences arising from the limited penetration by the field are ignored here

migrates equatorwards in intense storms. This is explicable in terms of a suggestion that precipated particles are guided by null lines in the field near the equatorial plane where the field of the ring current may locally cancel the ambient field. This suggestion also provides a qualitative explanation of the narrowness of the bands or arcs which are so striking a feature of well-developed aurorae. They frequently appear in the form of curtains waving across the sky with complex folds extending for thousands of kilometres more or less magnetic east-west. The vertical extent may be 100 km, from a height of 100 km upwards. The feature which requires a special explanation is their thinness; they may be only a few hundred metres thick. The auroral mechanism therefore provides either a guidance or particle selection along very narrow bands in the geomagnetic field and it seems possible that neutral lines in the field provide such guidance and that they wave about during a storm.

There may be several distinct auroral displays during a single storm, each one associated with a renewal of the auroral electrojet current system and presumably due to separate arrivals of vortices of solar plasma. During really great storms aurorae are observed well outside the accepted ranges of the auroral zones (magnetic latitudes 60°–70°), occasionally to latitudes as low as 30° (magnetic). This occurred several times during the International Geophysical Year (1957–58) when the network of observers was particularly effective at reporting occurrences of aurorae at widely distributed places. Aurorae are most likely to appear at middle or even low latitudes when intense storms follow one another so closely that their effects superimpose. Then a very strong ring current is built up, neutralizing the dipole field closer to the Earth than usual, causing precipitation of particles at lower latitudes.

Induced electric currents in the Earth

The electrical conductivity of the upper mantle is deduced by determining the contribution of induced currents in the earth to geomagnetic disturbances. The external and internal contributions to the disturbance field are mathematically separable by spherical harmonic analysis and, since the internal contribution is due to currents induced by the external contribution, some information about the conductivity of the Earth can be deduced. The upper mantle part of the conductivity profile in Figure 45 was estimated in this way. In that analysis spherical symmetry of the electrical conductivity in the mantle was assumed. However detailed examination of magnetic records shows differences between the records from different observatories which are incompatible with spherical symmetry and indicate horizontal heterogeneity of the upper mantle. Interest in mantle convection and global tectonics has stimulated a renewal of geomagnetic studies of conductivity, more explicitly directed to the search for high conductivity anomalies, which may be attributed to hot spots in the mantle. The principle by which the existence of a conductivity anomaly may be deduced from geomagnetic variations is indicated in Figure 60.

One of the problems which is limiting progress at present time is that sea-water is a sufficiently good conductor to modify geomagnetic disturbances and until its action is better understood geographic variations in disturbance vectors can only be unambiguously assigned to mantle conductivities if all of the observations are made at stations which are much closer to one another than they are to the nearest ocean. According to some authors the ocean effect has been overestimated. Results of measurements at continental margins then indicate that upper mantle conductivities are higher under oceans than under continents, i.e., the oceanic upper mantle is hotter. While this is clearly important to the theory of mantle convection, a final choice between rival explanations cannot yet be made and in fact neither appears entirely satisfactory. One is simply that hot rising limbs of mantle convective cells occur generally under oceans; the other considers convection to be irrelevant and higher temperatures to be due to a deeper distribution of radioactive elements. In the continents these are concentrated in the 30–40 km of crust and to suppose that the same amounts are distributed deeper under oceans is a simple way of explaining the equality of the geothermal flux from continents and oceans.

Further reading

The following articles give more detailed discussions of various aspects of geomagnetism as well as references to the original literature:

Stacey, F. D., *Physics of the Earth*, New York, Wiley, 1969
Matsushita, S. and Campbell, W. H. (editors), *Physics of Geomagnetic Phenomena* (2 vols.,), New York: Academic Press, 1967.
Irving, E., *Paleomagnetism*, New York: Wiley, 1964.
Williams, D. J. and Mead, G. D. (editors), *Magnetospheric Physics*, Washington: American Geophysical Union, 1969. Also in Reviews of Geophysics, Vol. 7, nos. 1–2

INDEX